PURE MATHEMATICS

8. VECTORS IN TWO AND THREE DIMENSIONS

Second Edition

By

Anthony Nicolaides

P.A.S.S. PUBLICATIONS

Private Academic & Scientific Studies Limited

© A. NICOLAIDES 1994, 2007

First Published in Great Britain 1994 by
Private Academic & Scientific Studies Limited

ISBN–13   978–1–872684–15–4

SECOND EDITION 2008

This book is copyright under the Berne Convention.
All rights are reserved. Apart as permitted under the Copyright Act, 1956, no part of this publication may be reproduced, stored in a retrieval system, or transmitted in any form of by any means, electronic, electrical, mechanical, optical, photocopying, recording or otherwise, without the prior permission of the publisher.

Titles by the same author.

Revised and Enhanced

1. Algebra.   GCE A Level   ISBN–13   978–1–872684–82–6   £11–95

2. Trigonometry.   GCE A Level   ISBN–13   978–1–872684–87–1   £11–95

3. Complex Numbers.   GCE A Level   ISBN–13   978–1–872684–92–5   £9–95

4. Differential Calculus and Applications.   GCE A Level   ISBN–13   978–1–872684–97–0   £9–95

5. Cartesian and Polar Curve Sketching.   GCE A Level   ISBN–13   978–1–872684–63–5   £9–95

6. Coordinate Geometry in two Dimensions.   GCE A Level   ISBN–13   978–1–872684–68–0   £9–95

7. Integral Calculus and Applications.   GCE A Level   ISBN–13   978–1–872684–73–4   £14–95

8. Vectors in two and three dimensions.   GCE A Level   ISBN–13   978–1–872684–15–4   £9–95

9. Determinants and Matrices.   GCE A Level   ISBN–13   978–1–872684–16–1   £9–95

10. Probabilities.   GCE A Level   ISBN–13   978–1–872684–17–8   £8–95
    This book includes the full solutions

11. Success in Pure Mathematics: The complete works of GCE A Level. (1–9 above inclusive)   ISBN–13   978–1–872684–93–2 £39–95

12. Electrical & Electronic Principles. First year Degree Level   ISBN–13   978–1–872684–98–7   £16–95

13. GCSE Mathematics Higher Tier Third Edition.   ISBN–13   978–1–872684–69–7   £19–95

All the books have answers and a CD is attached with FULL SOLUTIONS of all the exercises set at the end of the book.

Preface

This book, which is part of the GCE A level series in Pure Mathematics covers the specialized topic of Vectors.

The GCE A level series success in Pure Mathematics is comprised of nine books, covering the syllabuses of most examining boards. The books are designed to assist the student wishing to master the subject of Pure Mathematics. The series is easy to follow with minimum help. It can be easily adopted by a student who wishes to study it in the comforts of his home at his pace without having to attend classes formally; it is ideal for the working person who wishes to enhance his knowledge and qualification. The Vectors book, like all the books in the series, the theory is comprehensively dealt with, together with many worked examples and exercises. A step by step approach is adopted in all the worked examples. A CD is attached to the book with FULL SOLUTIONS of all the exercises set at the end of each chapter.

This book develops the basic concepts and skills that are essential for the GCE A level in Pure Mathematics.

The modules C_4, FP1, FP2, FP3 are dealt adequately in this book with ample examples.

A. Nicolaides

8. VECTORS IN TWO AND THREE DIMENSIONS
CONTENTS

1. INTRODUCTION TO VECTORS   1

Scalars   1

Vectors   1

Equal vectors (Direction and sense)   1

Magnitude of vectors |a| modulus   1

Triangle of forces   2

Free vectors   2

Position vectors   2

Algebra subtraction   2

The parallelogram of vectors   3

The polygon of vectors   3

Scalar multiple or submultiple of a vector   3

Exercises 1   5

2. VECTORS IN TWO AND THREE DIMENSIONS   **6**

Orthogonal vectors i, j, k   6

Modulus or magnitude (Unit vector)   6

Components of a vector in two and three dimensions   7

Direction ratios of a vector   9

Direction cosines of a vector   9

The position vector of the point T and PQ in the ratio $\lambda : \mu$ 12

Three dimensional coordinate geometry   12

Exercises 2   15

3. THE VECTOR EQUATION OF A STRAIGHT LINE   **17**

$\boxed{r = a + \lambda b}$ a = position vector, b = direction vector   17

The parametric equations of the line   17

The cartesian equation of the line   18

A line passing through two fixed points $\boxed{r = a + \lambda(b - a)}$ 18

Condition for three points to be collinear   19

Exercises 3   20

4. PAIRS OF LINES — 21

The intersection of a pair of lines l_1 and l_2 — 21

Parallel lines — 21

Skew line — 21

Angle between a pair of lines — 22

Exercises 4 — 23

5. COORDINATE GEOMETRY, IN 3 DIMENSIONS — 27

Equation of a plane — 27

The perpendicular distance of the point $A(x_1, y_1, z_1)$ from the plane $ax + by + cz = d$ is $\left| \dfrac{ax_1 + by_1 + cy_1 - d}{\sqrt{a^2 + b^2 + c^2}} \right|$ — 30

Equation of a plane in the form $\boxed{r \cdot n = d}$ — 30

Distance of a plane from the origin — 30

The distance between two parallel planes — 31

The parametric form of the vector equation of a plane $r = \lambda a + sb + tc$ — 31

plane passing through a given point and perpendicular to a given direction — 32

Perpendicular distance of a point from a plane — 32

The angle between two planes — 34

The angle between a line and a plane — 35

The intersection of two planes is a line — 35

The parametric form for the vector equation of a plane — 36

Exercises 5 — 37

6. VECTOR PRODUCT — 40

Introduction of vector product (cross product) — 40

The perpendicular distance, d, of a point $P(x_1, y_1, z_1)$ from a line with vector equation $r = a + \lambda b$ — 43

Area of triangle — 46

Volume of a tetrahedron — 48

Volume of a parallelepiped — 48

Triple scale product — 49

The shortest distance between two skew lines — 50

Formulae (summary) planes — 52

Exercises 6 — 53

Index — 55

1

Introduction to Vectors

Scalars

A scalar is a physical quantity that has only <u>magnitude</u>. The following are some examples of scalars. Length, area, volume, work done, electrical resistance, power, energy, mass, density, temperature, and potential. All the above physical quantities have a <u>magnitude</u>, the length is 15 cm, the electrical resistance is 5 Ω, the mass is 1.5 kg, the temperature is 23° C and so on.

Vectors

A vector is a physical quantity that has <u>magnitude and direction</u>. The following are some examples of vectors. Displacement, velocity, force, magnetic field strength, current and acceleration.

Simon Stevin of Bruges, a Mathematician and Engineer, in about 1586, was the first to show that a <u>force</u> can be represented by <u>a line, that is, a vector can by represented by a line</u>, and the addition of forces can be implemented using <u>triangle of forces</u>.

Equal Vectors

$\overrightarrow{AB}$ and $\overrightarrow{CD}$ are two vectors parallel and equal.

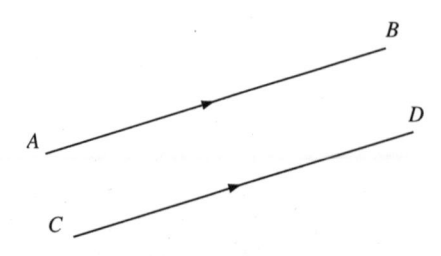

Fig. 8-I/1 Two equal vectors

Two vectors are equal if the following three properties are satisfied.

(1) The vectors are parallel.

(2) The vectors have the same <u>direction</u> and the same <u>sense</u>.

(3) The magnitude of vector, $\overrightarrow{AB}$, denoted by two vertical lines, $|\overrightarrow{AB}|$, is equal to the magnitude of vector $\overrightarrow{CD}$, $|\overrightarrow{CD}|$, that is,

$$|\overrightarrow{AB}| = |\overrightarrow{CD}|.$$

Note that there is a difference between the two words <u>direction</u> and <u>sense</u>.

The direction of the vector is from A to B and it is represented by $\overrightarrow{AB}$, or from B to A and is represented by $\overrightarrow{BA}$.

The sense of a vector could be clockwise or anticlockwise.

Consider three equal coplanar vectors $\overrightarrow{AB}$, $\overrightarrow{OP}$, and $\overrightarrow{CD}$. Since the three vectors are equal, we can represent them by the same algebraic symbol, **F**

$$\overrightarrow{AB} = \overrightarrow{CD} = \overrightarrow{OP} = \mathbf{F}.$$

The Magnitude of the Vectors are

$$|\overrightarrow{AB}| = |\overrightarrow{CD}| = |\overrightarrow{OP}| = |\mathbf{F}|.$$

The vertical lines denote the modulus or the magnitude of the vector and this is always considered to be positive.

1

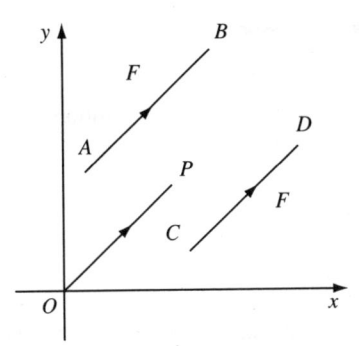

Fig. 8-I/2 Geometrical and algebraic representation of vectors

The word vector was first introduced by W. R. Hamilton, in 1844 in Dublin.

Triangle of Forces

Consider two forces to be applied at a point O at the directions and magnitudes shown in Fig. 8-I/3. It is required to find the total or <u>resultant force</u> applied at the point O.

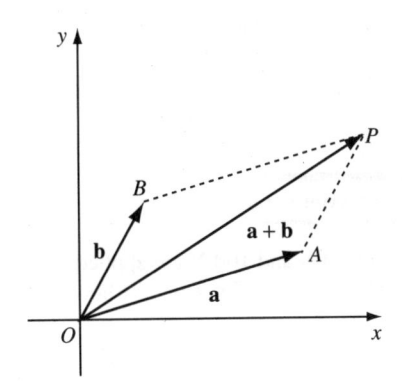

Fig. 8-I/3 Algebra addition of two vectors

Draw from B a line parallel and equal to $\overrightarrow{OA}$, $\overrightarrow{BP}$ and draw from A a line parallel and equal to $\overrightarrow{OB}$, $\overrightarrow{AP}$. The diagonal of the parallelogram $\overrightarrow{OP}$, is the resultant force $\mathbf{a} + \mathbf{b}$. But $\overrightarrow{AP} = \overrightarrow{OB} = \mathbf{b}$ and $\overrightarrow{OA} = \overrightarrow{BP} = \mathbf{a}$, therefore $\overrightarrow{OA} + \overrightarrow{AP} = \overrightarrow{OP}$ from the triangle of forces OAP. If $\overrightarrow{OA} = \mathbf{a}$, $\overrightarrow{OB} = \mathbf{b}$ and $\overrightarrow{OP} = \mathbf{c}$ then

$$\boxed{\mathbf{a} + \mathbf{b} = \mathbf{c}}$$

But also from the triangle of forces OBP, we have

$$\boxed{\mathbf{b} + \mathbf{a} = \mathbf{c}}$$

therefore, $\mathbf{a} + \mathbf{b} = \mathbf{b} + \mathbf{a}$ and the algebra addition is commutative.

Free Vectors

A <u>free vector</u> has no specified point of application. Fig. 8-I/3, shows two free vectors, $\overrightarrow{BP}$ and $\overrightarrow{AP}$.

Position Vectors

A position vector has a specified point of application, the origin O. Fig. 8-I/3 shows two position vectors, $\overrightarrow{OA}$ and $\overrightarrow{OB}$.

The origin of the vector, $\overrightarrow{OA}$, is O; similarly the origin of the vector, $\overrightarrow{OB}$, is O. The magnitudes of $\mathbf{a}$, $\mathbf{b}$ and $\mathbf{a} + \mathbf{b}$ are different and also the directions of $\mathbf{a}$, $\mathbf{b}$ and $\mathbf{a} + \mathbf{b}$ are different. The triangle of vectors OAP and OBP, $\overrightarrow{OA} + \overrightarrow{AP} = \overrightarrow{OP}$, $\overrightarrow{OB} + \overrightarrow{BP} = \overrightarrow{OP}$. The <u>sense</u> of the vectors $\overrightarrow{OA}$, $\overrightarrow{AP}$ is different from the sense of the vectors $\overrightarrow{OB}$ and $\overrightarrow{BP}$, the former sense is clockwise, the latter sense is anticlockwise.

Algebra Subtraction

If $\overrightarrow{OA} = \mathbf{a}$ in the direction shown in Fig. 8-I/4 then $\overrightarrow{OA'} = -\mathbf{a}$ is equal and opposite, $\overrightarrow{OA'}$ is a <u>negative vector</u>, $\overrightarrow{OA}$ and $\overrightarrow{OA'}$ have the same magnitudes and their directions are the same but of <u>opposite sense</u>.

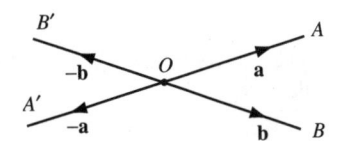

Fig. 8-I/4 Negative vectors

If $\overrightarrow{OA} = \mathbf{a}$ in the direction shown in Fig. 8-I/4 then $\overrightarrow{OA'} = -\mathbf{a}$ is equal and opposite to $\overrightarrow{OA}$, this defines a negative vector $(-\mathbf{a})$. Similarly for $-\mathbf{b}$. Completing the parallelogram $OBRA'$ as shown in Fig. 8-I/5 then $\overrightarrow{OR} = \mathbf{b} - \mathbf{a}$

$$\mathbf{b} - \mathbf{a} = \overrightarrow{OR}$$

$$-\mathbf{a} + \mathbf{b} = \overrightarrow{OR}$$

$$\boxed{\overrightarrow{OR} = \overrightarrow{OB} - \overrightarrow{OA'}}$$

Fig. 8-I/5 Vector subtraction

The Parallelogram of Vectors

The sum or resultant of two forces is found by using the parallelogram of forces.

The Polygon of Vectors

If we have more than two vectors, the polygon of vectors can be used to obtain the sum or resultant of vectors.

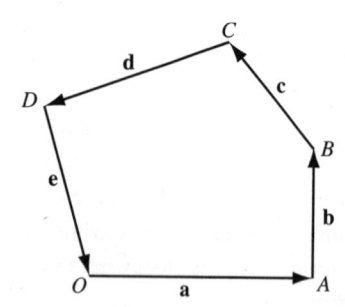

Fig. 8-I/6 The Polygon of vectors

$$\overrightarrow{OA} + \overrightarrow{AB} + \overrightarrow{BC} + \overrightarrow{CD} + \overrightarrow{DO} = 0$$

$$\mathbf{a} + \mathbf{b} + \mathbf{c} + \mathbf{d} + \mathbf{e} = 0.$$

Scalar Multiple or Submultiple of a Vector

If there are n equal vectors then $\mathbf{a} + \mathbf{a} + \ldots + \mathbf{a} = n\mathbf{a}$ where n is a scalar integer.

Examples

$$\mathbf{a} + \mathbf{a} + 2\mathbf{a} = 4\mathbf{a}$$

$$\mathbf{a} + \frac{1}{2}\mathbf{a} + \frac{3}{4}\mathbf{a} = \frac{4}{4}\mathbf{a} + \frac{2}{4}\mathbf{a} + \frac{3}{4}\mathbf{a} = \frac{9}{4}\mathbf{a}.$$

WORKED EXAMPLE 1

Distinguish clearly between a vector and a scalar physical quantities and state a few examples of each.

Solution 1

A vector is a physical quantity that possesses magnitude and direction such as displacement, velocity, acceleration, force, momentum, electric current, magnetic force, electric force.

A scalar is a physical quantity that possesses only magnitude such as mass, speed, temperature.

WORKED EXAMPLE 2

Distinguish clearly between position vector and free vector.

Solution 2

Position vector is a vector referred to a fixed point usually the origin of the axes. A position vector shows the displacement of one point relative to a fixed point O.

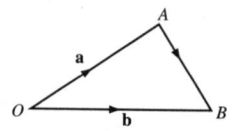

Fig. 8-I/7 Position vector

$\overrightarrow{OA} = \mathbf{a} =$ position vector

$\overrightarrow{OB} = \mathbf{b} =$ position vector

$\overrightarrow{AB} = \mathbf{b} - \mathbf{a} =$ free vector

= the displacement of point B relative to the point A, where A is not a fixed point.

The line of action of $\overrightarrow{AB}$ is different from $\overrightarrow{OA}$ and $\overrightarrow{OB}$ which both are referred to the fixed point O, the origin.

WORKED EXAMPLE 3

Two forces of 50 N and 100 N are applied at a point at the directions of 30° and 75° from the horizontal.

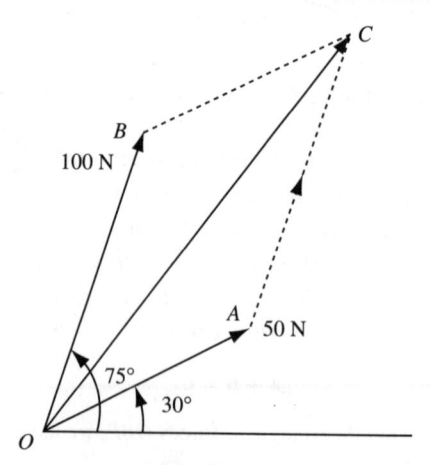

Fig. 8-I/8 Resultant of forces

Determine the resultant force and its direction from the horizontal.

Solution 3

This example can be solved graphically. Draw the line OA with a protractor at $30°$ to the horizontal, scale off along OA 5 cm, 1 cm = 10 N, from A draw the line AC, $75°$ to the horizontal, and scale off along AC, 10 cm, to represent 100 N, join O to C, $\overrightarrow{OC}$ is the resultant, measure its length and the angle that makes with the horizontal.

If a protractor and ruler are not available use calculations.

$\angle OAC = 135°$ and employing the cosine rule

$$OC^2 = 50^2 + 100^2 - 2 \times 50 \times 100 \times -0.7071$$

$$OC^2 = 2500 + 10000 + 7071$$

$$OC = 139.9 \approx 140 \text{ N the resultant force}$$

$$\frac{AC}{\sin \angle COA} = \frac{140}{\sin 135°} \Rightarrow \frac{100}{\sin \angle COA} = \frac{140}{0.7071}$$

$$\sin \angle COA = \frac{70.71}{140} = 0.505 \Rightarrow \angle COA = 30.34°$$

therefore the direction of the resultant from the horizontal is $30° + 30.34° = 60.3°$. Another idea of vectors consider the example.

Worked Example 4

A yacht is sailing at 10 km/h in a SE direction and a tide of 5 km/h is running towards the N, determine the velocity and direction of the yacht.

Solution 4

The bearings of the yacht sailing at 10 km/h and the tide of 5 km/h towards N are shown. $\left|\overrightarrow{OA}\right|$ is required and its direction.

$$OA^2 = 5^2 + 10^2 - 2 \times 5 \times 10 \cos 45°$$

$$= 25 + 100 - 100 \times 0.707$$

$$\overrightarrow{OA}^2 = 125 - 70.7 = 54.3 \Rightarrow \left|\overrightarrow{OA}\right| = 7.37 \text{ km/h}.$$

Using the sine rule

$$\frac{5}{\sin \angle AOB} = \frac{7.37}{\sin 45°}$$

$$\sin \angle AOB = \frac{5 \times \sin 45°}{7.37} = 0.4796472$$

$$\angle AOB = 28.66°.$$

The bearing of the yacht is $135° - 28.7° = 106.3°$.

Worked Example 5

A particle is moving around a circular orbit. Determine the magnitude and direction of the acceleration using vectors.

Solution 5

Consider a particle moving with constant speed v in a circle of radius r.

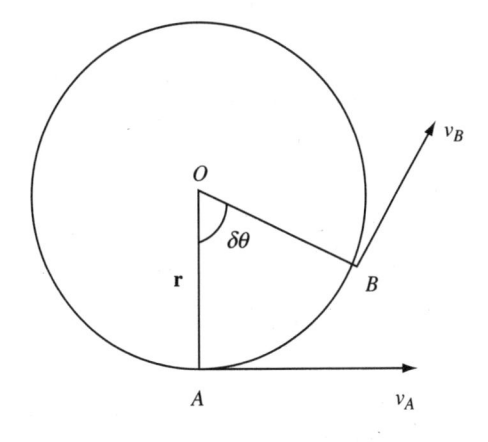

Fig. 8-I/10

A particle travels from A to B in a short interval of time δt, subtending an angle $\delta \theta$, as shown in Fig. 8-I/10

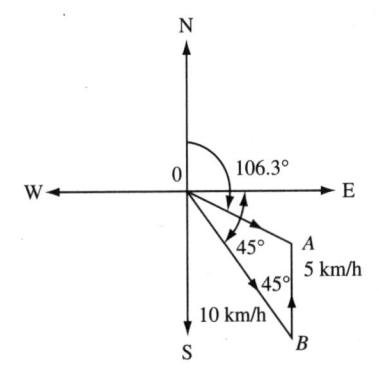

Fig. 8-I/9 The velocity of the yacht

speed = $\dfrac{\text{distance}}{\text{time}}$

$v = \dfrac{\text{arc } AB}{\delta t}$

$v = \dfrac{r\delta\theta}{\delta t}$.

Let v_A and v_B be the velocities at A an B respectively, these are tangential at A and B.

$a = $ acceleration $=$ change of velocity
$= v_B - v_A = v_B + (-v_A)$.

Drawing a vector PQ in the same direction as v_B and equal to v_B, QR in the opposite direction to v_A in the same direction as v_A and equal to v_A.

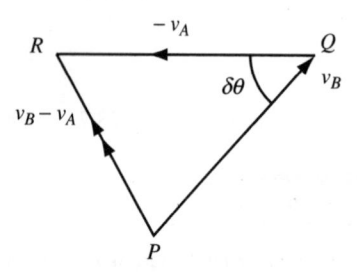

Fig. 8-I/11

PR is the resultant of the two velocities, the change of the angle between v_B and v_A is $\delta\theta$.

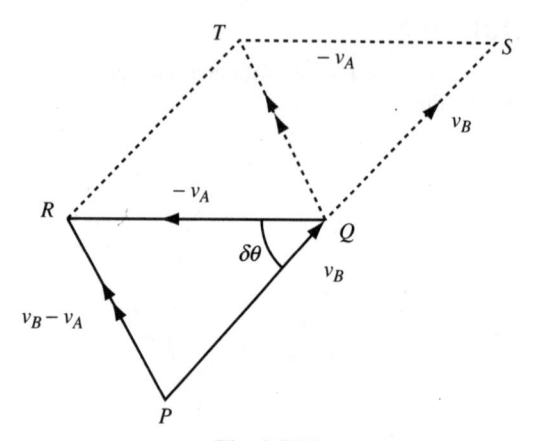

Fig. 8-I/12

Draw the parallelogram $QSTR$, PQ is the magnitude of v_B, RQ is the magnitude of $-v_A$, the resultant PR shows the magnitude and direction of acceleration, the change of velocity.

$PR = v\delta\theta = vv\dfrac{\delta t}{r}$

$a = \dfrac{\text{change in velocity}}{\text{time interval}} = \dfrac{PR}{\delta\theta} = \dfrac{v^2}{r}$

$$a = \dfrac{v^2}{r}$$

since $v = r\omega$, $\boxed{a = \omega^2 r}$

The direction of $\overrightarrow{PR}$ is towards the centre O of the circle and the acceleration is called <u>centripetal acceleration</u>. Therefore the magnitude of the acceleration is $\dfrac{v^2}{r}$ or $\omega^2 r$ and it is directed towards the centre. Therefore velocities and accelerations are vector quantities.

Exercises 1

1. Distinguish clearly between scalars and vectors.

2. State clearly the three properties that are satisfied for two equal vectors.

   Distinguish between direction and sense.

3. The vector algebra addition is commutative. Explain and illustrate this statement.

4. Distinguish between a Free Vector and a Position Vector. Illustrate by means of a diagram.

5. Distinguish between positive and negative vectors.

6. Determine the resultant of two vectors using the parallelogram of vectors. Two forces of 100 N and 250 N are applied at a point, O, at the directions of 45° and 60° from the horizontal. Determine the resultant force and its direction from the horizontal. Use sine and cosine rules.

7. The bearing of a yacht sailing 25 km/h is 045° and a tide of 5 km/h is running towards the E, determine the velocity and bearing of the yacht.

8. P divides AB in the ratio 2 : 3. Determine the position vector of the point P.

2

Vectors in Two and Three Dimensions

Orthogonal Vectors

Let **i**, **j**, and **k** be the unit vectors along the *x*-axis, *y*-axis and *z*-axis respectively

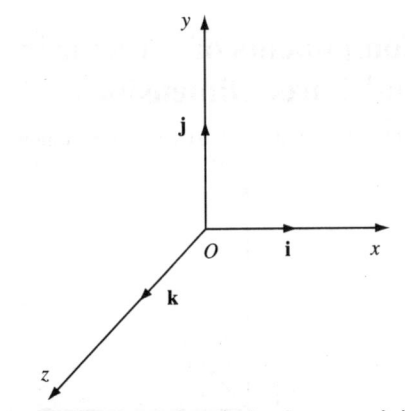

Fig. 8-I/13 The orthogonal unit vectors **i**, **j**, **k**.

x, *y* and *z* are three mutually perpendicular axes as shown in Fig. 8-I/13. Consider first for simplicity two dimensions.

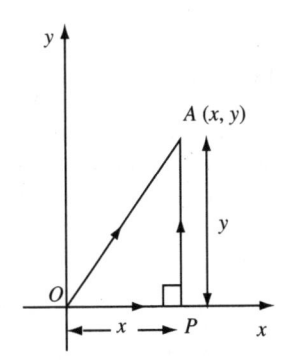

Fig. 8-I/14 The unit vectors **i**, **j**

Consider the position vector $\vec{OA} = x\mathbf{i} + y\mathbf{j}$ of the point A with coordinates, (x, y) and letting **i** and **j** be the unit vectors along the *x*-axis and *y*-axis respectively, then $\vec{OA} = x\mathbf{i} + y\mathbf{j}$, **i** and **j** denote the directions, therefore, we have x units along the *x*-axis, and y units along the *y*-axis.

In column matrix form $\vec{OA} = \begin{pmatrix} y \\ x \end{pmatrix} = x\mathbf{i} + y\mathbf{j}$.

The components of $\vec{OA}$ are $x\mathbf{i}$ and $y\mathbf{j}$, $\vec{OP} + \vec{PA} = \vec{OA}$

$\vec{OA} = x\mathbf{i} + y\mathbf{j}$.

Modulus or Magnitude in Two Dimensions

$$\left|\vec{OA}\right| = \sqrt{x^2 + y^2}$$

Extending the idea in three dimensions

$$\vec{OA} = x\mathbf{i} + y\mathbf{j} + z\mathbf{k}$$

$$= \begin{pmatrix} x \\ y \\ z \end{pmatrix}.$$

Modulus or magnitude in three dimension

$$\left|\vec{OA}\right| = \sqrt{x^2 + y^2 + z^2}$$

The unit vector.

A <u>unit vector</u> in a given direction is a vector with unit magnitude in that direction. Let **a** be a vector, its magnitude is denoted as |**a**| and its unit vector in the direction of **a** is denoted as **â** hat and defined

$$\hat{\mathbf{a}} = \frac{\mathbf{a}}{|\mathbf{a}|}$$

6

WORKED EXAMPLE 6

The following vectors are given:

(i) $\mathbf{a} = 3\mathbf{i} + 5\mathbf{j} - \mathbf{k}$
(ii) $\mathbf{b} = -\mathbf{i} - 2\mathbf{j} + 3\mathbf{k}$
(iii) $\mathbf{c} = 2\mathbf{i} + 3\mathbf{j} - 4\mathbf{k}$
(iv) $\mathbf{d} = -\mathbf{i} - \mathbf{j} - \mathbf{k}$
(v) $\mathbf{e} = \mathbf{i} + \mathbf{j} + \mathbf{k}$.

Determine the magnitudes of the vectors and hence find the corresponding unit vectors.

Solution 6

(i) $\mathbf{a} = 3\mathbf{i} + 5\mathbf{j} - \mathbf{k}$

$|\mathbf{a}|$ = the magnitude or modulus of the vector $\mathbf{a}$

$|\mathbf{a}| = \sqrt{(3)^2 + (5)^2 + (-1)^2}$

$= \sqrt{9 + 25 + 1} = \sqrt{35}$

$\hat{\mathbf{a}}$ = the unit vector in the direction of the vector $\mathbf{a}$.

$\hat{\mathbf{a}} = \dfrac{\mathbf{a}}{|\mathbf{a}|} = \dfrac{1}{\sqrt{35}} (3\mathbf{i} + 5\mathbf{j} - \mathbf{k})$

(ii) $\mathbf{b} = -\mathbf{i} - 2\mathbf{j} + 3\mathbf{k}$

$|\mathbf{b}| = \sqrt{(-1)^2 + (-2)^2 + 3^2}$

$= \sqrt{14}$

$\hat{\mathbf{b}} = \dfrac{\mathbf{b}}{|\mathbf{b}|} = \dfrac{1}{\sqrt{14}} (-\mathbf{i} - 2\mathbf{j} + 3\mathbf{k})$

= the unit vector in the direction of the vector $\mathbf{b}$.

(iii) $\mathbf{c} = 2\mathbf{i} + 3\mathbf{j} - 4\mathbf{k}$

$|\mathbf{c}| = \sqrt{2^2 + 3^2 + (-4)^2}$

$= \sqrt{29}$

$\hat{\mathbf{c}} = \dfrac{\mathbf{c}}{|\mathbf{c}|} = \dfrac{1}{\sqrt{29}} (2\mathbf{i} + 3\mathbf{j} - 4\mathbf{k})$

= the unit vector in the direction of the vector $\mathbf{c}$.

(iv) $\mathbf{d} = -\mathbf{i} - \mathbf{j} - \mathbf{k}$

$|\mathbf{d}| = \sqrt{(-1)^2 + (-1)^2 + (-1)^2} = \sqrt{3}$

$\hat{\mathbf{d}} = \dfrac{\mathbf{d}}{|\mathbf{d}|} = -\dfrac{1}{\sqrt{3}} (\mathbf{i} + \mathbf{j} + \mathbf{k})$

= the unit vector in the direction of the vector $\mathbf{d}$.

(v) $\mathbf{e} = \mathbf{i} + \mathbf{j} + \mathbf{k}$

$|\mathbf{e}| = \sqrt{1^2 + 1^2 + 1^2} = \sqrt{3}$

$\hat{\mathbf{e}} = \dfrac{1}{\sqrt{3}} (\mathbf{i} + \mathbf{j} + \mathbf{k})$

= the unit vector in the direction of the vector $\mathbf{e}$.

The Components of a Vector in Two and Three Dimensions

A vector $\overrightarrow{OA}$ can be expressed in two or three dimensions.

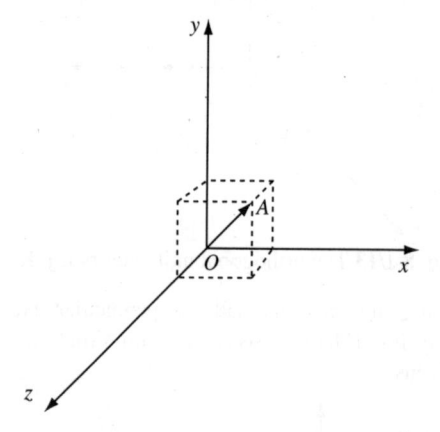

Fig. 8-I/15 The three components of $\overrightarrow{OA}$

If $\overrightarrow{OA}$ is the position vector of the point A with coordinates (x, y, z) and if $\mathbf{i}, \mathbf{j}$ and $\mathbf{k}$ are the unit vectors along the axes ox, oy, and oz respectively, then $\overrightarrow{OA} = x\mathbf{i} + y\mathbf{j} + z\mathbf{k}$ and therefore $x\mathbf{i}, y\mathbf{j}$ and $z\mathbf{k}$ are the components of the vector $\overrightarrow{OA}$ in the axes ox, oy and oz respectively.

$\overrightarrow{OA} = x\mathbf{i} + y\mathbf{j} + z\mathbf{k} = \begin{pmatrix} x \\ y \\ z \end{pmatrix}$

$\begin{pmatrix} x \\ y \\ z \end{pmatrix}$ is the corresponding column matrix.

Worked Example 7

Find the addition of the vectors $\mathbf{a} = 2\mathbf{i} + 3\mathbf{j} + 5\mathbf{k}$

$$\mathbf{b} = \mathbf{i} + 2\mathbf{j} + 3\mathbf{k}.$$

Solution 7

$\mathbf{a} + \mathbf{b} = (2\mathbf{i} + 3\mathbf{j} + 5\mathbf{k}) + (\mathbf{i} + 2\mathbf{j} + 3\mathbf{k}) = 3\mathbf{i} + 5\mathbf{j} + 8\mathbf{k}$

$$\mathbf{a} + \mathbf{b} = \begin{pmatrix} 2 \\ 3 \\ 5 \end{pmatrix} + \begin{pmatrix} 1 \\ 2 \\ 3 \end{pmatrix}$$

$$= \begin{pmatrix} 2+1 \\ 3+2 \\ 5+3 \end{pmatrix} = \begin{pmatrix} 3 \\ 5 \\ 8 \end{pmatrix}.$$

Worked Example 8

Determine the following operations

(i) $\mathbf{a} + 2\mathbf{b} + 3\mathbf{c}$

(ii) $3\mathbf{a} - \mathbf{b} + \mathbf{c}$

(iii) $-\mathbf{a} + 4\mathbf{b} - 2\mathbf{c}$.

If $\mathbf{a} = \mathbf{i} + 2\mathbf{j} - 3\mathbf{k}$, $\mathbf{b} = -3\mathbf{j} + 5\mathbf{k}$
and $\mathbf{c} = -4\mathbf{i} - \mathbf{j} + 2\mathbf{k}$.

Solution 8

(i) $\mathbf{a} + 2\mathbf{b} + 3\mathbf{c}$

$= (\mathbf{i} + 2\mathbf{j} - 3\mathbf{k}) + 2(-3\mathbf{j} + 5\mathbf{k})$

$\quad + 3(-4\mathbf{i} - \mathbf{j} + 2\mathbf{k})$

$= -11\mathbf{i} - 7\mathbf{j} + 13\mathbf{k}$

(ii) $3\mathbf{a} - \mathbf{b} + \mathbf{c}$

$= 3(\mathbf{i} + 2\mathbf{j} - 3\mathbf{k}) - (-3\mathbf{j} + 5\mathbf{k})$

$\quad + (-4\mathbf{i} - \mathbf{j} + 2\mathbf{k})$

$= -\mathbf{i} + 8\mathbf{j} - 12\mathbf{k}$

(iii) $-\mathbf{a} + 4\mathbf{b} - 2\mathbf{c}$

$= -(\mathbf{i} + 2\mathbf{j} - 3\mathbf{k}) + 4(-3\mathbf{j} + 5\mathbf{k})$

$\quad - 2(-4\mathbf{i} - \mathbf{j} + 2\mathbf{k})$

$= 7\mathbf{i} - 12\mathbf{j} + 19\mathbf{k}$.

Worked Example 9

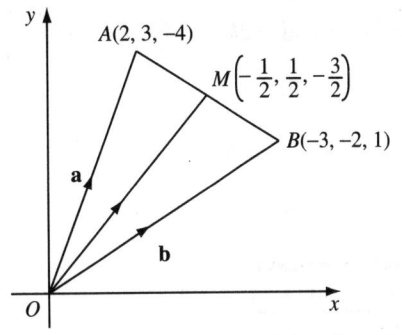

Fig. 8-I/16 Position vector of the mid-point

The position vectors of A and B are $\mathbf{a} = 2\mathbf{i} + 3\mathbf{j} - 4\mathbf{k}$ and $\mathbf{b} = -3\mathbf{i} - 2\mathbf{j} + \mathbf{k}$, find the position vector of the mid-point of AB. Determine the magnitude of $\overrightarrow{AB}$

Solution 9

The coordinates of A and B are $(2, 3, -4)$ and $(-3, -2, 1)$ respectively, the mid-point of AB is
$M\left(\dfrac{2-1}{2}, \dfrac{3-2}{2}, \dfrac{-4+1}{2}\right) \equiv M\left(-\dfrac{1}{2}, \dfrac{1}{2}, -\dfrac{3}{2}\right)$
hence the position vector of the mid-point is

$$\overrightarrow{OM} = -\frac{1}{2}\mathbf{i} + \frac{1}{2}\mathbf{j} - \frac{3}{2}\mathbf{k}.$$

$\overrightarrow{AB} = \mathbf{b} - \mathbf{a} = -3\mathbf{i} - 2\mathbf{j} + \mathbf{k} - 2\mathbf{i} - 3\mathbf{j} + 4\mathbf{k}$

$\quad = -5\mathbf{i} - 5\mathbf{j} + 5\mathbf{k}$

$\left|\overrightarrow{AB}\right| = \sqrt{(-5)^2 + (-5)^2 + 5^2} = \sqrt{75} = 5\sqrt{3}$.

Worked Example 10

The position vectors are given by the following coordinates

(i) $A(1, 2, 3)$

(ii) $B(-1, 2, -3)$

(iii) $C(0, 3, 5)$

(iv) $D(-4, 2, 1)$ and

(v) $E(3, 0, 4)$.

Write down the position vectors in the form $a\mathbf{i} + b\mathbf{j} + c\mathbf{k}$.

Solution 10

(i) $\vec{OA} = \mathbf{i} + 2\mathbf{j} + 3\mathbf{k}$

(ii) $\vec{OB} = -\mathbf{i} + 2\mathbf{j} - 3\mathbf{k}$

(iii) $\vec{OC} = 3\mathbf{j} + 5\mathbf{k}$

(iv) $\vec{OD} = -4\mathbf{i} + 2\mathbf{j} + \mathbf{k}$

(v) $\vec{OE} = 3\mathbf{i} + 4\mathbf{k}$.

Worked Example 11

Calculate the moduli of the vectors given:

(a) $\mathbf{u} = \mathbf{i} - 2\mathbf{j} + \sqrt{20}\,\mathbf{k}$

(b) $\mathbf{v} = -3\mathbf{i} + 7\mathbf{j} + 4\mathbf{k}$.

Solution 11

(a) $|\mathbf{u}| = \sqrt{1^2 + (-2)^2 + \left(\sqrt{20}\right)^2}$

$= \sqrt{1 + 4 + 20} = 5$

(b) $|\mathbf{v}| = \sqrt{(-3)^2 + 7^2 + 4^2}$

$= \sqrt{9 + 49 + 16} = \sqrt{74} = 8.60$ to 3 s.f.

Worked Example 12

Show that the points P, Q, R, S with position vectors $2\mathbf{j}$, $-2\mathbf{i}$, $-4\mathbf{j}$ and $2\mathbf{i} - 2\mathbf{j}$ respectively, are the vertices of a parallelogram.

Solution 12

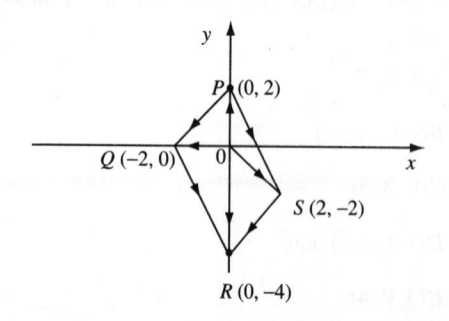

Fig. 8-I/17 Parallelogram

$\vec{PQ} = \vec{OQ} - \vec{OP} = -2\mathbf{j} - 2\mathbf{i}$,

$\vec{SR} = \vec{OR} - \vec{OS} = -4\mathbf{j} - 2\mathbf{i} + 2\mathbf{j}$

$= -2\mathbf{j} - 2\mathbf{i}$

$\vec{PQ} = \vec{SR} = -2\mathbf{j} - 2\mathbf{i}$

$\vec{QR} = \vec{OR} - \vec{OQ} = -4\mathbf{j} + 2\mathbf{i}$,

$\vec{PS} = \vec{OS} - \vec{OP} = 2\mathbf{i} - 2\mathbf{j} - 2\mathbf{j} = 2\mathbf{i} - 4\mathbf{j}$

$\vec{QR} = \vec{PS} = 2\mathbf{i} - 4\mathbf{j}$

$PQRS$ is a parallelogram.

Direction Ratios of a Vector

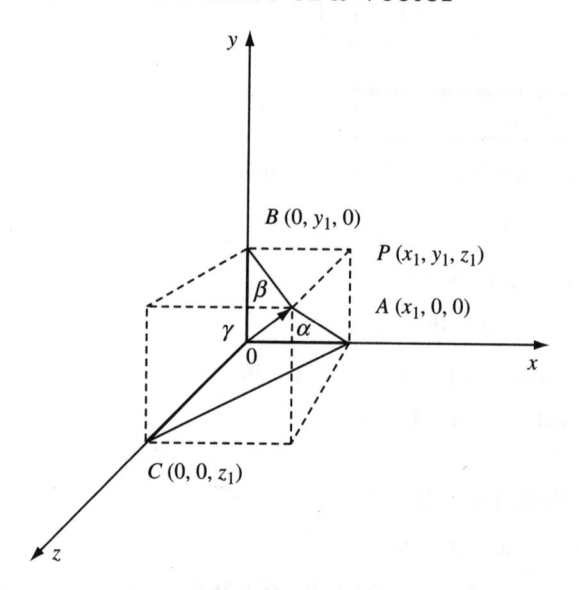

Fig. 8-I/18 Direction Ratios

Consider a position vector $\vec{OP} = x_1\mathbf{i} + y_1\mathbf{j} + z_1\mathbf{k}$.

The ratios $x_1 : y_1 : z_1$ are called the direction ratios of the vector $\vec{OP}$.

Direction Cosines of a Vector

The position vector $\vec{OP}$ makes three angles, α, β, γ with x-axis, y-axis and z-axis respectively as shown in the Fig. 8-I/18 above.

Connect P to A, B and C as shown $\angle PAO$, $\angle PBO$, $\angle PCO$ are 90° each as shown in Fig. 8-I/19 below.

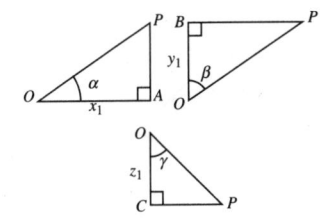

Fig. 8-I/19 Direction cosines

$$\cos a = \frac{x_1}{\left|\overrightarrow{OP}\right|} = l, \cos \beta = \frac{y_1}{\left|\overrightarrow{OP}\right|} = m,$$

$$\cos \gamma = \frac{z_1}{\left|\overrightarrow{OP}\right|} = n.$$

These cosines are called <u>direction cosines</u>.

Let $\left|\overrightarrow{OP}\right| = d$

$$\cos \alpha = \frac{x_1}{d} = l, \cos \beta = \frac{y_1}{d} = m,$$

$$\cos \gamma = \frac{z_1}{d} = n$$

$$\cos^2 \alpha + \cos^2 \beta + \cos^2 \gamma = \frac{x_1^2}{d^2} + \frac{y_1^2}{d^2} + \frac{z_1^2}{d^2}$$

$$= \frac{x_1^2 + y_1^2 + z_1^2}{d^2} = \frac{d^2}{d^2} = 1$$

but $\left|\overrightarrow{OP}\right| = d = \sqrt{x_1^2 + y_1^2 + z_1^2}$

$$\boxed{\cos^2 \alpha + \cos^2 \beta + \cos^2 \gamma = 1}$$

If the direction ratios of a line vector are $x_1 : y_1 : z_1$ then the direction cosines of this vector are $l : m : n$.

$$\boxed{l^2 + m^2 + n^2 = 1}$$

If $\overrightarrow{OP}$ is a position vector, $\overrightarrow{OP} = x_1\mathbf{i} + y_1\mathbf{j} + z_1\mathbf{k} = dl\mathbf{i} + dm\mathbf{j} + dn\mathbf{k} = d(l\mathbf{i} + m\mathbf{j} + n\mathbf{k})$ since d is a scalar quantity then the vector $l\mathbf{i} + m\mathbf{j} + n\mathbf{k}$ is a unit vector.

$\mathbf{a} = |\mathbf{a}|\hat{\mathbf{a}}$

$$\hat{\mathbf{a}} = \frac{\mathbf{a}}{|\mathbf{a}|} = \frac{x_1\mathbf{i} + y_1\mathbf{j} + z_1\mathbf{k}}{\sqrt{x_1^2 + y_1^2 + z_1^2}} = \frac{x_1}{d}\mathbf{i} + \frac{y_1}{d}\mathbf{j} + \frac{z_1}{d}\mathbf{k}$$

$$= l\mathbf{i} + m\mathbf{j} + n\mathbf{k}.$$

WORKED EXAMPLE 13

The following position vectors are given by the sets of coordinates $P(-1, 2, 4)$, $Q(-2, -3, -5)$, $R(7, 8, 9)$. Write down the vectors in the form $x\mathbf{i} + y\mathbf{j} + z\mathbf{k}$ and hence find the direction ratios and direction cosines.

Solution 13

$\overrightarrow{OP} = -\mathbf{i} + 2\mathbf{j} + 4\mathbf{k}$, $\overrightarrow{OQ} = -2\mathbf{i} - 3\mathbf{j} - 5\mathbf{k}$, $\overrightarrow{OR} = 7\mathbf{i} + 8\mathbf{j} + 9\mathbf{k}$ the corresponding direction ratios are $-1 : 2 : 4, -2 : -3 : -5, 7 : 8 : 9$.

The magnitudes of the vectors are:

$$\left|\overrightarrow{OP}\right| = \sqrt{(-1)^2 + 2^2 + 4^2}$$
$$= \sqrt{21}$$

$$\left|\overrightarrow{OQ}\right| = \sqrt{(-2)^2 + (-3)^2 + (-5)^2}$$
$$= \sqrt{38}$$

$$\left|\overrightarrow{OR}\right| = \sqrt{7^2 + 8^2 + 9^2} = \sqrt{194}$$

the direction cosines are correspondingly

$$\frac{-1}{\sqrt{21}} : \frac{2}{\sqrt{21}} : \frac{4}{\sqrt{21}}, \frac{-2}{\sqrt{38}} : \frac{-3}{\sqrt{38}} : \frac{-5}{\sqrt{38}},$$

$$\frac{7}{\sqrt{194}} : \frac{8}{\sqrt{194}} : \frac{9}{\sqrt{194}}.$$

WORKED EXAMPLE 14

Determine the direction cosines of the vector $\overrightarrow{OP} = \mathbf{i} + 2\mathbf{j} + 3\mathbf{k}$ and hence calculate the angles that $\overrightarrow{OP}$ make with the axes.

Solution 14

$1 : 2 : 3$ are the direction ratios

$$\left|\overrightarrow{OP}\right| = \sqrt{1^2 + 2^2 + 3^2} = \sqrt{14}$$

$$\frac{1}{\sqrt{14}} : \frac{2}{\sqrt{14}} : \frac{3}{\sqrt{14}} \text{ are the direction cosines}$$

$\cos \alpha : \cos \beta : \cos \gamma$

$\cos \alpha = \dfrac{1}{\sqrt{14}}$, hence $\alpha = 74.5°$

$\cos \beta = \dfrac{2}{\sqrt{14}}$, hence $\beta = 57.7°$

$\cos \gamma = \dfrac{3}{\sqrt{14}}$, hence $\gamma = 36.7°$

WORKED EXAMPLE 15

A position vector makes angles of 30° and 60° with the x-axis and z-axis respectively, determine the angle that the vector makes with the y-axis.

Solution 15

$$\cos^2 \alpha + \cos^2 \beta + \cos^2 \gamma = 1$$
$$\cos^2 30° + \cos^2 \beta + \cos^2 60° = 1$$
$$\cos^2 \beta = 1 - 0.75 - 0.25$$
$$\cos^2 \beta = 0$$
$$\boxed{\beta = 90°}$$

WORKED EXAMPLE 16

The position vectors $\overrightarrow{OP} = 3\mathbf{i} - 2\mathbf{j} - \mathbf{k}$ and $\overrightarrow{OQ}$ are referred to the origin O. Determine $\overrightarrow{OQ}$ if the free vector $\overrightarrow{PQ} = -2\mathbf{i} + \mathbf{j} + 4\mathbf{k}$.

Calculate

(i) the magnitudes of the vectors $\overrightarrow{OP}$, $\overrightarrow{OQ}$, and $\overrightarrow{PQ}$.

(ii) the units vectors corresponding to $\overrightarrow{OP}$, $\overrightarrow{OQ}$, and $\overrightarrow{PQ}$

(iii) the direction ratios of the vectors $\overrightarrow{OP}$, $\overrightarrow{OQ}$, and $\overrightarrow{PQ}$

(iv) the directions cosines of the vectors and hence the angles with respect to x-axis, y-axis and z-axis.

Solution 16

$$\overrightarrow{OP} + \overrightarrow{PQ} = \overrightarrow{OQ}$$
$$3\mathbf{i} - 2\mathbf{j} - \mathbf{k} + (-2\mathbf{i} + \mathbf{j} + 4\mathbf{k}) = \overrightarrow{OQ}$$
$$\boxed{\overrightarrow{OQ} = \mathbf{i} - \mathbf{j} + 3\mathbf{k}}$$

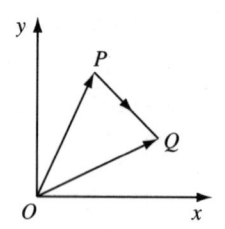

Fig. 8-I/20 Position vector

(i) $\left|\overrightarrow{OP}\right| = \sqrt{3^2 + (-2)^2 + (-1)^2} = \sqrt{14}$

$\left|\overrightarrow{OQ}\right| = \sqrt{1^2 + (-1)^2 + 3^2} = \sqrt{11}$

$\left|\overrightarrow{PQ}\right| = \sqrt{(-2)^2 + 1^2 + 4^2} = \sqrt{21}$

(ii) $\overrightarrow{OP} = \left|\overrightarrow{OP}\right| O\hat{P}$

$O\hat{P} = \dfrac{\overrightarrow{OP}}{\left|\overrightarrow{OP}\right|} = \dfrac{3}{\sqrt{14}}\mathbf{i} - \dfrac{2}{\sqrt{14}}\mathbf{j} - \dfrac{1}{\sqrt{14}}\mathbf{k}$

$\overrightarrow{OQ} = \left|\overrightarrow{OQ}\right| O\hat{Q}$

$O\hat{Q} = \dfrac{\overrightarrow{OQ}}{\left|\overrightarrow{OQ}\right|} = \dfrac{1}{\sqrt{11}}\mathbf{i} - \dfrac{1}{\sqrt{11}}\mathbf{j} + \dfrac{3}{\sqrt{11}}\mathbf{k}$

$\overrightarrow{PQ} = \left|\overrightarrow{PQ}\right| P\hat{Q}$

$P\hat{Q} = \dfrac{\overrightarrow{PQ}}{\left|\overrightarrow{PQ}\right|} = -\dfrac{2}{\sqrt{21}}\mathbf{i} + \dfrac{1}{\sqrt{21}}\mathbf{j} + \dfrac{4}{\sqrt{21}}\mathbf{k}$

(iii) $l:m:n$ are the direction ratios.

$3:-2:-1$ for $\overrightarrow{OP}$

$1:-1:3$ for $\overrightarrow{OQ}$

$-2:1:4$ for $\overrightarrow{PQ}$

(iv) $\cos \alpha$, $\cos \beta$ and $\cos \gamma$ are direction cosines

$\dfrac{3}{\sqrt{14}} : -\dfrac{2}{\sqrt{14}} : -\dfrac{1}{\sqrt{14}}$ for $\overrightarrow{OP}$

$\cos \alpha = \dfrac{3}{\sqrt{14}}$, $\cos \beta = -\dfrac{2}{\sqrt{14}}$, $\cos \gamma = -\dfrac{1}{\sqrt{14}}$

$\alpha = 36.7°$, $\beta = 122.3°$, $\gamma = 105.5°$

$\dfrac{1}{\sqrt{11}} : -\dfrac{1}{\sqrt{11}} : \dfrac{3}{\sqrt{11}}$ for $\overrightarrow{OQ}$

$\cos \alpha = \frac{1}{\sqrt{11}}, \cos \beta = -\frac{1}{\sqrt{11}},$

$\cos \gamma = \frac{3}{\sqrt{11}}$

$\alpha = 72.5°, \beta = 107.6°, \gamma = 25.2°$

$-\frac{2}{\sqrt{21}} : \frac{1}{\sqrt{21}} : \frac{4}{\sqrt{21}}$ for $\overrightarrow{PQ}$

$\cos \alpha = -\frac{2}{\sqrt{21}}, \cos \beta = \frac{1}{\sqrt{21}},$

$\cos \gamma = \frac{4}{\sqrt{21}}$

$\alpha = 115.9°, \beta = 77.4°, = 29.2°.$

The Position Vector of the Point T Dividing PQ in the Ratio $\lambda : \mu$

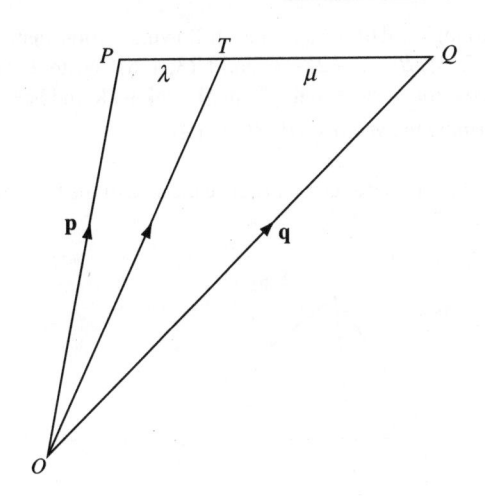

Fig. 8-I/21 Position vector of a point dividing PQ in the ratio $\lambda : \mu$

where **p** and **q** are the position vectors of the points P and Q.

$\dfrac{\overrightarrow{PT}}{\overrightarrow{TQ}} = \dfrac{\lambda}{\mu}$

$\dfrac{\overrightarrow{PT}}{\overrightarrow{PQ}} = \dfrac{\lambda}{\lambda + \mu}$

$\overrightarrow{PT} = \dfrac{\lambda}{\lambda + \mu} \overrightarrow{PQ}$

$\overrightarrow{OT} = \overrightarrow{OP} + \overrightarrow{PT}$

$= \mathbf{p} + \dfrac{\lambda}{\lambda + \mu} (\mathbf{q} - \mathbf{p})$

$= \dfrac{\mathbf{p}(\lambda + \mu) + \lambda(\mathbf{q} - \mathbf{p})}{\lambda + \mu}$

$= \dfrac{\mathbf{p}\lambda + \mathbf{p}\mu + \mathbf{q}\lambda - \mathbf{p}\lambda}{\lambda + \mu}$

$\boxed{\overrightarrow{OT} = \dfrac{\mathbf{p}\mu + \mathbf{q}\lambda}{\lambda + \mu}}$

Three Dimension Coordinate Geometry

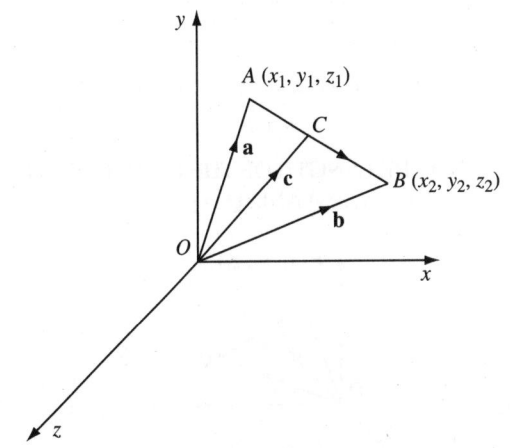

Fig. 8-I/22

$\overrightarrow{OA} = x_1 \mathbf{i} + y_1 \mathbf{j} + z_1 \mathbf{k}$

the position vector of $A(x_1, y_1, z_1)$

$\overrightarrow{OB} = x_2 \mathbf{i} + y_2 \mathbf{j} + z_2 \mathbf{k}$

the position vector of $B(x_2, y_2, z_2)$

$\overrightarrow{AB} = \overrightarrow{OB} - \overrightarrow{OA}$ the free vector

$= (x_2 - x_1) \mathbf{i} + (y_2 - y_1) \mathbf{j} + (z_2 - z_1) \mathbf{k}$

LENGTH OF $\overrightarrow{AB}$

$\left| \overrightarrow{AB} \right| = \sqrt{(x_2 - x_1)^2 + (y_2 - y_1)^2 + (z_2 - z_1)^2}$

DIRECTION RATIOS OF $\overrightarrow{AB}$

$(x_2 - x_1) : (y_2 - y_1) : (z_2 - z_1)$

DIRECTION COSINES OF $\vec{AB}$

$$\frac{x_2 - x_1}{|\vec{AB}|} : \frac{y_2 - y_1}{|\vec{AB}|} : \frac{z_2 - z_1}{|\vec{AB}|}$$

POINT DIVIDING AB in the ratio $\lambda : \mu$

Let C divide the line AB internally in the ratio $\lambda : \mu$, then $\dfrac{\vec{AC}}{\vec{CB}} = \dfrac{\lambda}{\mu}$

$$\vec{OC} = \vec{OA} + \vec{AC} = \vec{OA} + \frac{\lambda}{\lambda + \mu}\vec{AB}$$

$$= \vec{OA} + \frac{\lambda}{\lambda + \mu}(\mathbf{b} - \mathbf{a}) = \mathbf{a} + \frac{\lambda}{\lambda + \mu}(\mathbf{b} - \mathbf{a})$$

$$= \frac{(\lambda + \mu)\mathbf{a} + \lambda \mathbf{b} - \lambda \mathbf{a}}{\lambda + \mu} = \frac{\mu \mathbf{a} + \lambda \mathbf{b}}{\lambda + \mu}$$

$$= \frac{\mu(x_1 \mathbf{i} + y_1 \mathbf{j} + z_1 \mathbf{k}) + \lambda(x_2 \mathbf{i} + y_2 \mathbf{j} + z_2 \mathbf{k})}{\lambda + \mu}$$

TO FIND THE LENGTH OF THE LINE JOINING $P(x_1, y_1, z_1)$ AND $Q(x_2, y_2, z_2)$

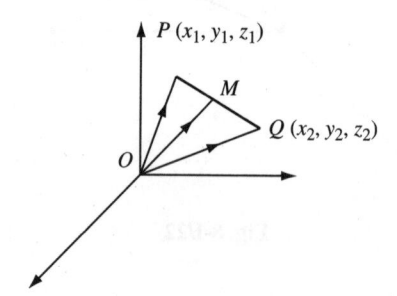

Fig. 8-I/23

The position vectors $\vec{OP}$ and $\vec{OQ}$ are

$\vec{OP} = x_1 \mathbf{i} + y_1 \mathbf{j} + z_1 \mathbf{k}$ and

$\vec{OQ} = x_2 \mathbf{i} + y_2 \mathbf{j} + z_2 \mathbf{k}$.

$\vec{PQ} = \vec{OQ} - \vec{OP}$

$\quad = (x_2 - x_1)\mathbf{i} + (y_2 - y_1)\mathbf{j} + (z_2 - z_1)\mathbf{k}$

$|\vec{PQ}| = \sqrt{(x_2 - x_1)^2 + (y_2 - y_1)^2 + (z_2 - z_1)^2}$

TO FIND THE MIDPOINT OF PQ

Let M be the mid-point of PQ, the position vector

$\vec{OM} = \vec{OP} + \vec{PM}$

$\quad = \vec{OP} + \dfrac{1}{2}\vec{PQ}$

$\quad = \vec{OP} + \dfrac{1}{2}\left(\vec{OQ} - \vec{OP}\right)$

$\quad = \dfrac{1}{2}\vec{OP} + \dfrac{1}{2}\vec{OQ}$

$$\boxed{\vec{OM} = \dfrac{1}{2}\left(\vec{OP} + \vec{OQ}\right)}$$

$\vec{OM} = \dfrac{1}{2}[(x_1 + x_2)\mathbf{i} + (y_1 + y_2)\mathbf{j} + (z_1 + z_2)\mathbf{k}]$

$M\left(\dfrac{1}{2}(x_1 + x_2), \dfrac{1}{2}(y_1 + y_2), \dfrac{1}{2}(z_1 + z_2)\right)$

WORKED EXAMPLE 17

A triangle ABC has the following coordinates: $A(1, 2, 3)$, $B(-1, -2, -3)$ and $C(3, 4, 6)$. Write down the position vectors in the form $a\mathbf{i} + b\mathbf{j} + c\mathbf{k}$ and hence determine the vectors $\vec{AB}$, $\vec{BC}$ and $\vec{AC}$.

Calculate the perimeter and hence the area of the triangle ABC.

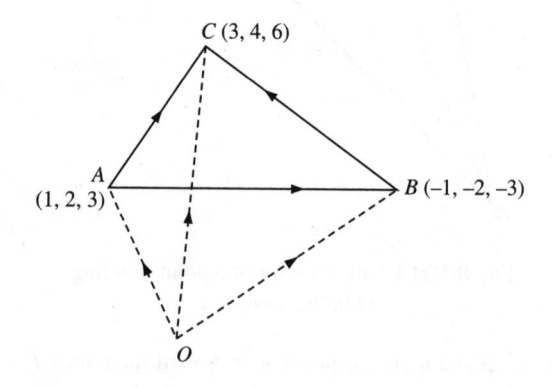

Fig. 8-I/24 The perimeter and the area of $\triangle ABC$

Solution 17

$\vec{AB} = \vec{OB} - \vec{OA}$

$\quad = (-\mathbf{i} - 2\mathbf{j} - 3\mathbf{k}) - (\mathbf{i} + 2\mathbf{j} + 3\mathbf{k})$

$\quad = -2\mathbf{i} - 4\mathbf{j} - 6\mathbf{k}$

$$\vec{BC} = \vec{OC} - \vec{OB}$$
$$= (3\mathbf{i} + 4\mathbf{j} + 6\mathbf{k}) - (-\mathbf{i} - 2\mathbf{j} - 3\mathbf{k})$$
$$= 4\mathbf{i} + 6\mathbf{j} + 9\mathbf{k}$$

$$\vec{AC} = \vec{OC} - \vec{OA}$$
$$= (3\mathbf{i} + 4\mathbf{j} + 6\mathbf{k}) - (\mathbf{i} + 2\mathbf{j} + 3\mathbf{k})$$
$$= 2\mathbf{i} + 2\mathbf{j} + 3\mathbf{k}$$

$$\vec{AB} + \vec{BC} = \vec{AC}$$

L.H.S. $= \vec{AB} + \vec{BC}$
$$= (-2\mathbf{i} - 4\mathbf{j} - 6\mathbf{k}) + (4\mathbf{i} + 6\mathbf{j} + 9\mathbf{k})$$
$$= 2\mathbf{i} + 2\mathbf{j} + 3\mathbf{k}$$

$$\left|\vec{AB}\right| = \sqrt{(-2)^2 + (-4)^2 + (-6)^2}$$
$$= \sqrt{4 + 16 + 36}$$
$$= \sqrt{56} = 7.48$$

$$\left|\vec{BC}\right| = \sqrt{4^2 + 6^2 + 9^2}$$
$$= \sqrt{16 + 36 + 81}$$
$$= \sqrt{133} = 11.5$$

$$\left|\vec{AC}\right| = \sqrt{2^2 + 2^2 + 3^2}$$
$$= \sqrt{17} = 4.12.$$

Each preceding and subsequent calculation is worked out to 3 significant figures.

The perimeter $= 7.48 + 11.5 + 4.12 = 23.1$

$s =$ semi-perimeter
$$= \frac{23.1}{2} = 11.6 \text{ units.}$$

Area $\triangle = \sqrt{s(s-a)(s-b)(s-c)}$
$$= \sqrt{\begin{array}{c}11.6 \times (11.6 - 7.48)(11.6 - 11.5) \\ (11.6 - 4.12)\end{array}}$$
$$= \sqrt{11.6 \times 4.12 \times 0.1 \times 7.48}$$
$$= 5.98 \text{ square units.}$$

If we round off only at the end the answer would be 3.6 s.u.

WORKED EXAMPLE 18

The position vectors of the triangle PQR are
$$\vec{OP} = 2\mathbf{i} + 5\mathbf{j} - 7\mathbf{k}$$
$$\vec{OQ} = -3\mathbf{i} - 2\mathbf{j} + 5\mathbf{k}$$
$$\vec{OR} = -\mathbf{i} + 3\mathbf{j} - 6\mathbf{k}.$$

Determine the vectors of the sides of the triangle and hence calculate the perimeter and the area of the triangle.

Solution 18

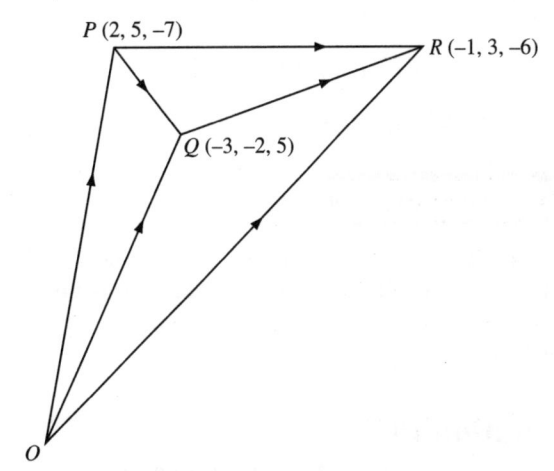

Fig. 8-I/25 Perimeter and area of $\triangle PQR$

$$\vec{PQ} = \vec{OQ} - \vec{OP}$$
$$= (-3\mathbf{i} - 2\mathbf{j} + 5\mathbf{k}) - (2\mathbf{i} + 5\mathbf{j} - 7\mathbf{k})$$
$$= -5\mathbf{i} - 7\mathbf{j} + 12\mathbf{k}$$

$$\vec{QR} = \vec{OR} - \vec{OQ}$$
$$= (-\mathbf{i} + 3\mathbf{j} - 6\mathbf{k}) - (-3\mathbf{i} - 2\mathbf{j} + 5\mathbf{k})$$
$$= 2\mathbf{i} + 5\mathbf{j} - 11\mathbf{k}$$

$$\vec{PR} = \vec{OR} - \vec{OP} = (-\mathbf{i} + 3\mathbf{j} - 6\mathbf{k}) - (2\mathbf{i} + 5\mathbf{j} - 7\mathbf{k})$$
$$= -3\mathbf{i} - 2\mathbf{j} + \mathbf{k}$$

Each subsequent calculation is worked out to three significant figures.

$$\left|\vec{PQ}\right| = \sqrt{(-5)^2 + (-7)^2 + (12)^2}$$
$$= \sqrt{25 + 49 + 144} = \sqrt{218} = 14.8$$

$$|\overrightarrow{QR}| = \sqrt{2^2 + 5^2 + (-11)^2} = \sqrt{4 + 25 + 121}$$
$$= \sqrt{150} = 12.2$$

$$|\overrightarrow{PR}| = \sqrt{(-3)^2 + (-2)^2 + 1^2} = \sqrt{9 + 4 + 1}$$
$$= \sqrt{14} = 3.74$$

the semi-perimeter $s = \dfrac{14.8 + 12.2 + 3.74}{2} = 15.4$

hence $2s = 30.8$ units.

Area $\triangle = \sqrt{s(s-a)(s-b)(s-c)}$
$$= \sqrt{15.4 \times 0.6 \times 3.2 \times 11.66}$$
$$= 18.3 \text{ square units.}$$

WORKED EXAMPLE 19

The position vectors of the points A and B are $-2\mathbf{i} + 4\mathbf{j} - 3\mathbf{k}$ and $3\mathbf{i} - 5\mathbf{j} + 2\mathbf{k}$ respectively. Find the position vector of the point P which divides AB internally in the ratio 3:2.

Solution 19

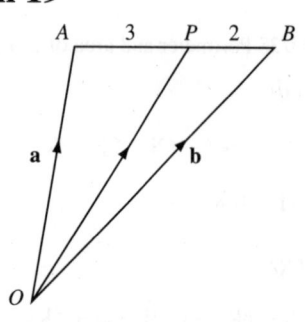

Fig. 8-I/26 Position vector $\overrightarrow{OP}$

$$\overrightarrow{OP} = \frac{\mathbf{a}\mu + \mathbf{b}\lambda}{\lambda + \mu} = \frac{\mathbf{a}2 + \mathbf{b}3}{3 + 2}$$
$$= \frac{2(-2\mathbf{i} + 4\mathbf{j} - 3\mathbf{k}) + 3(3\mathbf{i} - 5\mathbf{j} + 2\mathbf{k})}{5}$$
$$= \frac{-4\mathbf{i} + 8\mathbf{j} + -6\mathbf{k} + 9\mathbf{i} - 15\mathbf{j} + 6\mathbf{k}}{5}$$
$$= \frac{5\mathbf{i} - 7\mathbf{j}}{5} = \mathbf{i} - \frac{7}{5}\mathbf{j}.$$

Exercises 2

1. The position vectors are given by the following coordinates
   (i) $A(1, 2, 3)$
   (ii) $B(-1, 2, -3)$
   (iii) $C(0, 3, 5)$
   (iv) $D(-4, 2, 1)$ and
   (v) $E(3, 0, 4)$.

   Write down the vectors in the form $a\mathbf{i} + b\mathbf{j} + c\mathbf{k}$.

2. The position vectors of A, B and C are given $2\mathbf{i} + 3\mathbf{j} - \mathbf{k}$, $-\mathbf{i} + 2\mathbf{j} - 4\mathbf{k}$, $-3\mathbf{i} + \mathbf{j} + \mathbf{k}$ respectively. Write down the coordinates of A, B and C.

3. Calculate the modulus of the vectors given:
   (a) $\mathbf{u} = \mathbf{i} - 2\mathbf{j} + \sqrt{20}\mathbf{k}$
   (b) $\mathbf{v} = -3\mathbf{i} + 7\mathbf{j} + 4\mathbf{k}$.

4. Determine the magnitude of the lines $\overrightarrow{OP}, \overrightarrow{OQ}, \overrightarrow{OR}$ where $P(3, 4, 5), Q(-2, -1, 1), R(2, -3, 5)$.

5. If $\mathbf{a} = \begin{pmatrix} 1 \\ 2 \\ 3 \end{pmatrix}, \mathbf{b} = \begin{pmatrix} 2 \\ 2 \\ 2 \end{pmatrix}, \mathbf{c} = \begin{pmatrix} 3 \\ 0 \\ 5 \end{pmatrix},$

   find
   (i) $|\mathbf{a}|$
   (ii) $|\mathbf{b} - \mathbf{a}|$
   (iii) $\left|2\mathbf{c} - \dfrac{1}{2}\mathbf{a}\right|$.

6. The position vectors $\overrightarrow{OP}, \overrightarrow{OQ},$ and $\overrightarrow{OR}$ are given by $P(-1, -2, -3), Q(1, 4, 7), R(3, -5, 8)$. Determine $\overrightarrow{PQ}, \overrightarrow{PR}$ and $\overrightarrow{QR}$ and hence find their moduli.

7. Determine the direction ratios and direction cosines for the following vectors:
   (i) $\mathbf{u} = \mathbf{i} - 3\mathbf{j} + 5\mathbf{k}$
   (ii) $\mathbf{v} = -2\mathbf{i} + 4\mathbf{j} - 6\mathbf{k}$
   (iii) $\mathbf{w} = 3\mathbf{i} - 7\mathbf{j} + 11\mathbf{k}$.

8. The direction cosines of a vector are $\dfrac{5}{7} : \dfrac{4}{7} : \dfrac{2\sqrt{2}}{7}$. Calculate the angles that this vector is making with x, y, and z axes respectively. Check that $\cos^2 \alpha + \cos^2 \beta + \cos^2 \gamma = 1$.

9. The angles that a vector make with the x and y axes are 45° and 67.5° respectively. Calculate the angle that it makes with the z-axis.

10. Show that $\cos^2 \alpha + \cos^2 \beta + \cos^2 \gamma = 1$.

11. Show that the position vectors of a point P dividing a line AB in the ratio $\lambda : \mu$ where **a** and **b** are the position vectors of the points P and Q is given by

$$\overrightarrow{OP} = \frac{\mathbf{a}\mu + \mathbf{b}\lambda}{\lambda + \mu}$$

12. The direction cosines of a vector are $\dfrac{2\sqrt{10}}{9} : \dfrac{5}{9} : \dfrac{4}{9}$. Calculate the angles that this vector is making with x, y and z respectively.

13. Deduce that the points with position vectors **a**, **b** and $\mathbf{a}\mu + \mathbf{b}\lambda$ are collinear provided $\lambda + \mu = 1$.

14. Find the direction cosines of the line joining (a, b, c) to (l, m, n).

15. Find the direction cosines of the line joining $(1, 2, 3)$ to $(3, -4, 5)$.

16. If the direction cosines of a line are in the ratios $3 : 7 : 11$ find the actual direction cosines of the line.

17. A line makes angles of 70° and 80° with the positive directions of the x-axis and z-axis respectively. Find the angle it makes with the positive direction of the y-axis.

18. Find the direction cosines of a line that makes equal angles with the three axes.

19. If $\gamma = 2\alpha$ and $\beta = \alpha$. Find the direction cosines.

20. A triangle ABC has the following position vectors:

$$\overrightarrow{OA} = 2\mathbf{i} - 3\mathbf{j} + \mathbf{k}$$
$$\overrightarrow{OB} = \mathbf{i} + 5\mathbf{k}$$
$$\overrightarrow{OC} = -\mathbf{i} + 5\mathbf{j} + 7\mathbf{k}$$

Determine the free vectors $\overrightarrow{AB}$, $\overrightarrow{BC}$ and $\overrightarrow{AC}$. Calculate the perimeter ABC and hence determine the area of the triangle using Heron's Formula.

3

The Vector Equation of a Straight Line

$r = a + \lambda b$

The vector equation through a fixed point A, of position vector $\mathbf{a} = a_1\mathbf{i} + a_2\mathbf{j} + a_3\mathbf{k}$ and parallel to a vector $\mathbf{b} = b_1\mathbf{i} + b_2\mathbf{j} + b_3\mathbf{k}$.

Let $\overrightarrow{OA}$ be the position vector given as $\mathbf{a} = a_1\mathbf{i} + a_2\mathbf{j} + a_3\mathbf{k}$ and $\mathbf{b} = b_1\mathbf{i} + b_2\mathbf{j} + b_3\mathbf{k}$. It is required to find the vector equation of a straight line passing through A and parallel to a vector $\mathbf{b}$.

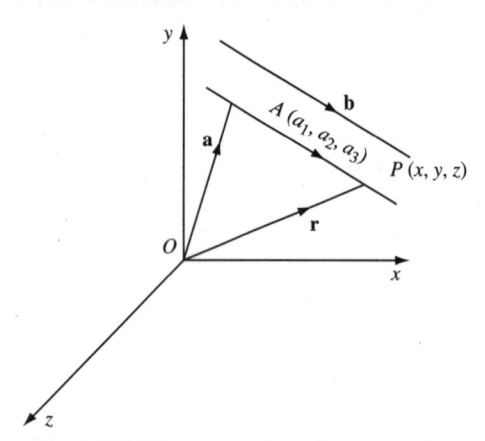

Fig. 8-I/27 Vector equation of a straight line

Let $P(x, y, z)$ be a point on the straight line whose vector equation is required.

$\overrightarrow{AP} = \lambda \mathbf{b}$ where λ is a scalar parameter. From the triangle OAP we have

$\overrightarrow{OA} + \overrightarrow{AP} = \overrightarrow{OP}$, the clockwise sum of the vectors is equal to the anticlockwise vector.

$\overrightarrow{OA} = \mathbf{a} = a_1\mathbf{i} + a_2\mathbf{j} + a_3\mathbf{k}$ and $\overrightarrow{AP} = \lambda\mathbf{b} = \lambda(b_1\mathbf{i} + b_2\mathbf{j} + b_3\mathbf{k})$

$\mathbf{a} + \lambda\mathbf{b} = \mathbf{r}$

$$\boxed{\mathbf{r} = \mathbf{a} + \lambda\mathbf{b}}$$

WORKED EXAMPLE 20

Determine the vector equation of the line which passes through the point $A(-3, 2, 4)$ and is parallel to the vector $\mathbf{b} = \mathbf{i} - \mathbf{j} + 3\mathbf{k}$.

Solution 20

Let the vector equation of the line be $\mathbf{r} = \mathbf{a} + \lambda\mathbf{b}$ where $\mathbf{a} = -3\mathbf{i} + 2\mathbf{j} + 4\mathbf{k}$ and $\mathbf{b} = \mathbf{i} - \mathbf{j} + 3\mathbf{k}$.

$$\mathbf{r} = (-3\mathbf{i} + 2\mathbf{j} + 4\mathbf{k}) + \lambda(\mathbf{i} - \mathbf{j} + 3\mathbf{k})$$
$$\mathbf{r} = (-3 + \lambda)\mathbf{i} + (2 - \lambda)\mathbf{j} + (4 + 3\lambda)\mathbf{k}$$

is the vector equation, which is the position vector of $P(x, y, z)$.

The Parametric Equations of the Line

If the position vector of the line is given $\mathbf{r} = x\mathbf{i} + y\mathbf{j} + z\mathbf{k}$ then

$$\mathbf{r} = x\mathbf{i} + y\mathbf{j} + z\mathbf{k} = (a_1\mathbf{i} + a_2\mathbf{j} + a_3\mathbf{k}) \\ + \lambda(b_1\mathbf{i} + b_2\mathbf{j} + b_3\mathbf{k})$$

$$x\mathbf{i} + y\mathbf{j} + z\mathbf{k} = (a_1 + b_1\lambda)\mathbf{i} + (a_2 + b_2\lambda)\mathbf{j} \\ + (a_3 + b_3\lambda)\mathbf{k}$$

Equating the coefficients of $\mathbf{i}, \mathbf{j}$ and $\mathbf{k}$ we have $x = a_1 + b_1\lambda$, $y = a_2 + b_2\lambda$, $z = a_3 + b_3\lambda$ which are the parametric equations of the line.

WORKED EXAMPLE 21

Find the parametric equation of the vector equation of the line

$$\mathbf{r} = (-3 + \lambda)\mathbf{i} + (2 - \lambda)\mathbf{j} + (4 + 3\lambda)\mathbf{k}.$$

Solution 21

$\mathbf{r} = x\mathbf{i} + y\mathbf{j} + z\mathbf{k} = (-3+\lambda)\mathbf{i} + (2-\lambda)\mathbf{j} + (4+3\lambda)\mathbf{k}$.
Equating the coefficients of $\mathbf{i}$, $\mathbf{j}$ and $\mathbf{k}$ we have

$$x = -3 + \lambda$$
$$y = 2 - \lambda$$
$$z = 4 + 3\lambda$$

the parametric equations of the line.

The Cartesian Equation of the Line

The parametric equations of the line are $x = a_1 + b_1\lambda$, $y = a_2 + b_2\lambda$, $z = a_3 + b_3\lambda$ and

$$x = a_1 + b_1\lambda \text{ or } \frac{x - a_1}{b_1} = \lambda$$
$$y = a_2 + b_2\lambda \text{ or } \frac{y - a_2}{b_2} = \lambda$$
$$z = a_3 + b_3\lambda \text{ or } \frac{z - a_3}{b_3} = \lambda.$$

The cartesian equations of the line are $\frac{x - a_1}{b_1} = \frac{y - a_2}{b_2} = \frac{z - a_3}{b_3} = \lambda$ $b_1 : b_2 : b_3$ are the direction ratios of the line.

WORKED EXAMPLE 22

Determine the cartesian equations of the line whose parametric equations are $x = -3 + \lambda$, $y = 2 - \lambda$, and $z = 4 + 3\lambda$ and hence find the direction ratios of the line.

Solution 22

$x = -3 + \lambda \quad \text{or} \quad \frac{x+3}{1} = \lambda$

$y = 2 - \lambda \quad \text{or} \quad \frac{-2+y}{-1} = \lambda$

$z = 4 + 3\lambda \quad \text{or} \quad \frac{z-4}{3} = \lambda$

$\lambda = \frac{x+3}{1} = \frac{-2+y}{-1} = \frac{z-4}{3}.$

The direction ratios of the line are $1 : -1 : 3$.

WORKED EXAMPLE 23

A line passes through a fixed point $A(4, 5, -7)$ and is parallel to a vector $2\mathbf{i} - 3\mathbf{j} + 5\mathbf{k}$. Determine the following:

(i) The vector equation of the line.
(ii) The parametric equations of the line.
(iii) The cartesian equations of the line.

Solution 23

(i) $\mathbf{r} = \mathbf{a} + \lambda\mathbf{b}$ the vector equation of the line,

$\mathbf{r} = (4\mathbf{i} + 5\mathbf{j} - 7\mathbf{k}) + \lambda(2\mathbf{i} - 3\mathbf{j} + 5\mathbf{k})$ or
$\mathbf{r} = (4 + 2\lambda)\mathbf{i} + (5 - 3\lambda)\mathbf{j} + (-7 + 5\lambda)\mathbf{k}$.

(ii) The parametric equations of the line
$\mathbf{r} = x\mathbf{i} + y\mathbf{j} + z\mathbf{k}$
$= (4 + 2\lambda)\mathbf{i} + (5 - 3\lambda)\mathbf{j} + (-7 + 5\lambda)\mathbf{k}$

equating the coefficients of $\mathbf{i}$, $\mathbf{j}$ and $\mathbf{k}$

$x = 4 + 2\lambda$ ...(1)
$y = 5 - 3\lambda$ ...(2)
$z = -7 + 5\lambda$ ...(3)

the parametric equations of the line.

(iii) The cartesian equations of the line are found by equating λ in each case of (1), (2) and (3)
$\lambda = \frac{x-4}{2} = \frac{y-5}{-3} = \frac{z+7}{5}.$

A Line Passing through Two Fixed Points

Let the position vectors of the two fixed points be $\overrightarrow{OA} = x_1\mathbf{i} + y_1\mathbf{j} + z_1\mathbf{k} = \mathbf{a}$ and $\overrightarrow{OB} = x_2\mathbf{i} + y_2\mathbf{j} + z_2\mathbf{k} = \mathbf{b}$ respectively.

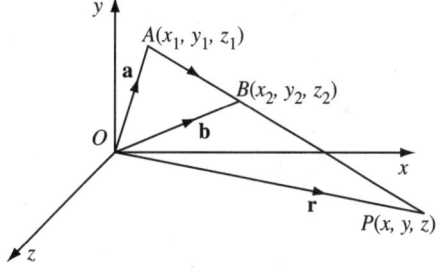

Fig. 8-I/28 Vector equation of a line passing through two fixed points

Let P be a point on AB produced $\overrightarrow{OA} + \overrightarrow{AB} = \overrightarrow{OB}$ from which we have

$\overrightarrow{AB} = \overrightarrow{OB} - \overrightarrow{OA}$

$\overrightarrow{AB} = \mathbf{b} - \mathbf{a}$

$\overrightarrow{AP} = t\overrightarrow{AB} = t(\mathbf{b} - \mathbf{a}).$

From OAP triangle, we have

$$\mathbf{r} = \overrightarrow{OA} + \overrightarrow{AP} = \mathbf{a} + t(\mathbf{b} - \mathbf{a})$$

$$\boxed{\mathbf{r} = \mathbf{a} + t(\mathbf{b} - \mathbf{a}).}$$

The vector equation of the line which passes through two fixed points $A(x_1, y_1, z_1)$ and $B(x_2, y_2, z_2)$ whose position vectors are $\mathbf{a}$ and $\mathbf{b}$ respectively.
This vector equation of the line can be expressed as column matrices

$$\boxed{\mathbf{r} = x\mathbf{i} + y\mathbf{j} + z\mathbf{k} = \begin{pmatrix} x_1 \\ y_1 \\ z_1 \end{pmatrix} + t \begin{pmatrix} x_2 - x_1 \\ y_2 - y_1 \\ z_2 - z_1 \end{pmatrix}}$$

or

$$\boxed{\begin{aligned} \mathbf{r} &= x\mathbf{i} + y\mathbf{j} + z\mathbf{k} \\ &= (x_1\mathbf{i} + y_1\mathbf{j} + z_1\mathbf{k}) + t[(x_2 - x_1)\mathbf{i} + (y_2 - y_1)\mathbf{j} \\ &\quad + (z_2 - z_1)\mathbf{k}] \end{aligned}}$$

Condition for Three Points to be Collinear

Let the points A, B and C be collinear as shown in the diagram and let the position vectors be $\mathbf{a}$, $\mathbf{b}$ and $\mathbf{c}$ respectively. The equation of the line BC is given $\mathbf{r} = \mathbf{b} + \lambda(\mathbf{c} - \mathbf{b})$ where $\mathbf{b}$ is the position vector and $\mathbf{c} - \mathbf{b}$ is the direction vector and λ is a scalar parameter.

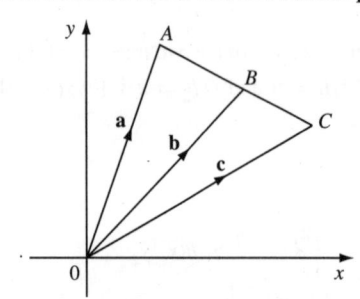

Fig. 8-I/29 Collinear points

The position vector of A must satisfy the equation of BC, substituting $\mathbf{r} = \mathbf{a}$

$$\mathbf{a} = \mathbf{b} + \lambda(\mathbf{c} - \mathbf{b})$$
$$\mathbf{a} + \lambda\mathbf{b} - \mathbf{b} - \lambda\mathbf{c} = 0$$
$$\mathbf{a} + \mathbf{b}(\lambda - 1) - \lambda\mathbf{c} = 0$$

this can be expressed as

$$\mathbf{a} + s\mathbf{b} + t\mathbf{c} = 0$$
$$1 + s + t = 0$$
$$s = -t - 1$$

or $t = -s - 1$.

The Vector Equation of a Straight Line — 19

WORKED EXAMPLE 24

Prove that the points $A(1, -1, 0)$, $B(2, 0, 3)$, $C(0, -2, -3)$ are collinear.

Solution 24

The position vectors of B and C are $\overrightarrow{OB} = 2\mathbf{i} + 3\mathbf{k}$, $\overrightarrow{OC} = -2\mathbf{j} - 3\mathbf{k}$ and the equation of the line BC is given

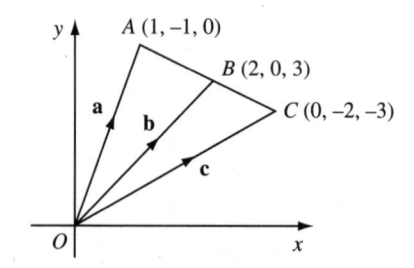

Fig. 8-I/30 Collinear points

$$\mathbf{r} = \mathbf{b} + \lambda(\mathbf{c} - \mathbf{b})$$
$$\mathbf{r} = 2\mathbf{i} + 3\mathbf{k} + \lambda(-2\mathbf{j} - 3\mathbf{k} - 2\mathbf{i} - 3\mathbf{k})$$
$$\mathbf{r} = 2\mathbf{i} + 3\mathbf{k} + \lambda(-2\mathbf{i} - 2\mathbf{j} - 6\mathbf{k})$$
$$\mathbf{r} = (2 - 2\lambda)\mathbf{i} - 2\lambda\mathbf{j} + (3 - 6\lambda)\mathbf{k}.$$

If A is on the line BC then

$$\mathbf{r} = \mathbf{i} - \mathbf{j}$$
$$\mathbf{i} - \mathbf{j} = (2 - 2\lambda)\mathbf{i} - 2\lambda\mathbf{j} + (3 - 6\lambda)\mathbf{k}$$

equating the coefficients of $\mathbf{i}$, $\mathbf{j}$ and $\mathbf{k}$ we have

$$2 - 2\lambda = 1, \quad \lambda = \frac{1}{2}$$
$$-2\lambda = -1, \quad \lambda = \frac{1}{2}$$
$$3 - 6\lambda = 0, \quad \lambda = \frac{1}{2}.$$

WORKED EXAMPLE 25

Three points A, B and C are collinear, the position vectors of A and C are given, $\overrightarrow{OA} = 2\mathbf{i} - 3\mathbf{j} + 4\mathbf{k}$, $\overrightarrow{OC} = -3\mathbf{i} + 2\mathbf{j} + 5\mathbf{k}$, find the position vector of B, for $\lambda = -1$.

Solution 25

The equation of the line on AC is given $\mathbf{r} = \mathbf{a} + \lambda(\mathbf{c} - \mathbf{a})$ since B lies on AC then

$x_1\mathbf{i} + y_1\mathbf{j} + z_1\mathbf{k} = \mathbf{a} + \lambda(\mathbf{c} - \mathbf{a})$

$\qquad = (2\mathbf{i} - 3\mathbf{j} + 4\mathbf{k}) + \lambda(-3\mathbf{i} + 2\mathbf{j}$
$\qquad\qquad + 5\mathbf{k} - 2\mathbf{i} + 3\mathbf{j} - 4\mathbf{k})$

$\qquad = (2\mathbf{i} - 3\mathbf{j} + 4\mathbf{k}) + \lambda(-5\mathbf{i} + 5\mathbf{j} + \mathbf{k})$

$\qquad = (2 - 5\lambda)\mathbf{i} + (5\lambda - 3)\mathbf{j} + (4 + \lambda)\mathbf{k}$

$x_1 = 2 - 5\lambda = 2 - 5(-1) = 7$

$y_1 = 5\lambda - 3 = 5(-1) - 3 = -8$

$z_1 = 4 + \lambda = 4 - 1 = 3$

therefore the position vector of B is

$\boxed{\overrightarrow{OB} = 7\mathbf{i} - 8\mathbf{j} + 3\mathbf{k}.}$

Exercises 3

1. Determine the position and direction vectors for the following:

   (a) $\dfrac{x+1}{1} = \dfrac{y+2}{2} = \dfrac{z+3}{3} = \lambda$

   (b) $\dfrac{x-1}{2} = \dfrac{y-2}{-3} = \dfrac{z-3}{4} = t$

   (c) $\dfrac{x}{3} = \dfrac{y+3}{-4} = \dfrac{z-1}{2} = \mu$

   (d) $\dfrac{x+a}{p} = \dfrac{y+b}{q} = \dfrac{z+c}{r} = s.$

2. Find the direction ratios and hence the direction cosines of the vectors of question 1.

3. The line $\dfrac{x+1}{1} = \dfrac{y+2}{2} = \dfrac{z+3}{3}$ is parallel to the line $\dfrac{x+a}{l} = \dfrac{y+b}{m} = \dfrac{z+c}{n}$ which passes through the point $(-3, -4, -5)$. Deduce the values of $a, b, c, l, m,$ and n.

4. The line $\dfrac{x+a}{-3} = \dfrac{y+b}{-4} = \dfrac{z+c}{-5}$ is parallel to the line $\dfrac{x-1}{l} = \dfrac{y-2}{m} = \dfrac{z-3}{n}$ and passes through the point $(2, 2, 2)$. Deduce the values of $a, b, c, l, m,$ and n.

5. Determine the unit vectors in the direction of the lines of question 1.

6. Find the vector equation of the line which passes through the point $A(-1, 2, 5)$ and is parallel to the direction vector $2\mathbf{i} - 3\mathbf{j} + 7\mathbf{k}$.

7. Find the vector equation of the line which passes through the point $\mathbf{a} = \mathbf{i} - \mathbf{j} + 3\mathbf{k}$ and is parallel to the direction vector $\mathbf{b} = 2\mathbf{i} + 3\mathbf{j} - 5\mathbf{k}$.

8. Find the vector equation of the line which passes through a point with position vector $\mathbf{a} = a_1\mathbf{i} + a_2\mathbf{j} + a_3\mathbf{k}$ and is parallel to the direction vector $\mathbf{b} = b_1\mathbf{i} + b_2\mathbf{j} + b_3\mathbf{k}$.

9. Find the vector equation of the line which passes through the points $A(1, -2, 3)$ and $B(-2, 4, 7)$.

10. Find the vector equation of the line which passes through two points with position vectors $\mathbf{w}$ and $\mathbf{u}$.

11. Find the vector equation of the line which passes through two points with position vector $\mathbf{a} = 2\mathbf{i} + 2\mathbf{j} + 2\mathbf{k}$, and $\mathbf{b} = 3\mathbf{i} + 3\mathbf{j} + 3\mathbf{k}$.

12. Explain the difference of the vector equations of a line

    $\mathbf{r} = \mathbf{a} + \lambda\mathbf{b} \qquad \mathbf{r} = \mathbf{a} + \lambda(\mathbf{b} - \mathbf{a}).$

13. Write the following vector equations

    $\mathbf{r}_1 = (\mathbf{i} - 3\mathbf{j} + 2\mathbf{k}) + \lambda(-2\mathbf{i} + 3\mathbf{j} - 4\mathbf{k})$

    $\mathbf{r}_2 = (2\mathbf{i} + \mathbf{j} - \mathbf{k}) + \mu(\mathbf{i} + 7\mathbf{j} - \mathbf{k})$

    $\mathbf{r}_3 = (\mathbf{i} + \mathbf{j} + \mathbf{k}) + \nu(2\mathbf{i} + 2\mathbf{j} - 3\mathbf{k})$

    in column vector forms.

14. The vector equations of three lines are given

    (a) $\mathbf{r}_1 = (3\mathbf{j} - 5\mathbf{k}) + t(2\mathbf{i} - 3\mathbf{j} + 7\mathbf{k})$
    (b) $\mathbf{r}_2 = (\mathbf{i} - \mathbf{j} + 7\mathbf{k}) + \mu(\mathbf{i} + 3\mathbf{j} + 4\mathbf{k})$
    (c) $\mathbf{r}_3 = (-2\mathbf{i} + \mathbf{j} - 4\mathbf{k}) + \lambda(-3\mathbf{i} + 5\mathbf{j} + \mathbf{k}).$

    Find three points, A, B and C on each line for three values of the parameters

    (a) $t = 1, 2, -3$ (b) $\mu = 0, -2, 1$
    (c) $\lambda = -2, -1, 4.$

15. Show that the point with position vector $\mathbf{i} + 5\mathbf{j} + 13\mathbf{k}$ lies on the line l with vector equation

    $\mathbf{r} = 5\mathbf{i} - \mathbf{j} + 7\mathbf{k} + \lambda(-2\mathbf{i} + 3\mathbf{j} + 3\mathbf{k}).$

16. Show that the point with position vector $3\mathbf{i} - 2\mathbf{j} + 4\mathbf{k}$ lies on the line l with vector equation

    $\mathbf{r} = (5\mathbf{j} + 7\mathbf{k}) + \mu(3\mathbf{i} - 7\mathbf{j} - 3\mathbf{k}).$

17. A line passes through the point $(-2, 4, -5)$ and is parallel to the direction vector $3\mathbf{i} + 5\mathbf{j} + \mathbf{k}$. Three points on the line are $A(-2, 4, -5)$, $B(-5, -1, -6)$, $C(4, 14, -3)$. Find the values of the parameters for these points.

4

Pairs of Lines

The vector equations of a pair of lines are given as $\mathbf{r}_1 = \mathbf{a}_1 + \lambda \mathbf{b}_1$ and $\mathbf{r}_2 = \mathbf{a}_2 + \mu \mathbf{b}_2$ where $\mathbf{a}_1 = x_1\mathbf{i} + y_1\mathbf{j} + z_1\mathbf{k}$, $\mathbf{a}_2 = x_2\mathbf{i} + y_2\mathbf{j} + z_2\mathbf{k}$ are fixed points and the lines are parallel to $\mathbf{b}_1 = x_1'\mathbf{i} + y_1'\mathbf{j} + z_1'\mathbf{k}$ and $\mathbf{b}_2 = x_2'\mathbf{i} + y_2'\mathbf{j} + z_2'\mathbf{k}$ respectively.

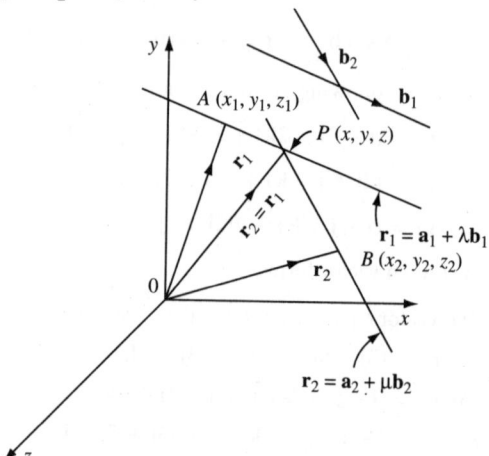

Fig. 8-I/31 Pairs of lines

The line $l_1 : \mathbf{r}_1 = \mathbf{a}_1 + \lambda \mathbf{b}_1$, passes through the fixed point A and it is parallel to the vector $\mathbf{b}_1$.

The line $l_2 : \mathbf{r}_2 = \mathbf{a}_2 + \mu \mathbf{b}_2$, passes through the fixed point B and it is parallel to the vector $\mathbf{b}_2$.

The direction vectors $\mathbf{b}_1$ and $\mathbf{b}_2$ may be parallel, intersect or neither (skew vectors). The most important case to consider in details is the intersection of the line l_1, and l_2 or the intersection of the direction $\mathbf{b}_1$ and $\mathbf{b}_2$.

The Intersection of a Pair of Lines l_1 and l_2

Fig. 8-I/31 shows the case of the two lines l_1 and l_2 that intersect at a point P with coordinates (x, y, z), the vector equations are equal at this point

$\mathbf{r}_1 = \mathbf{r}_2 \quad \mathbf{a}_1 + \lambda \mathbf{b}_1 = \mathbf{a}_2 + \mu \mathbf{b}_2$

where λ and μ are such parameters that satisfy this equality.

Parallel Lines

The direction vectors b_1 and b_2 are parallel.

Skew Lines

The direction vectros b_1 and b_2 are neither parallel nor intersect.

WORKED EXAMPLE 26

The following pairs of lines are given

(i) $l_1 : r_1 = (3\mathbf{i} + \mathbf{j} - 4\mathbf{k}) + \lambda(2\mathbf{i} - 3\mathbf{j} + \mathbf{k})$
  $l_2 : r_2 = (-2\mathbf{i} + 4\mathbf{j} + \mathbf{k}) + \mu(-\mathbf{i} + 4\mathbf{j} - 7\mathbf{k})$

(ii) $l_1 : r_1 = (-2\mathbf{i} - 2\mathbf{j} - 4\mathbf{k}) + \lambda(-\mathbf{i} + 2\mathbf{j} + 3\mathbf{k})$
  $l_2 : r_2 = (\mathbf{i} + \mathbf{j} + \mathbf{k}) + \mu(\mathbf{i} - 7\mathbf{j} + 2\mathbf{k})$

(iii) $l_1 : r_1 = (8\mathbf{i} - 8\mathbf{j} + 18\mathbf{k}) + \lambda(2\mathbf{i} + 2\mathbf{j} + 6\mathbf{k})$
  $l_2 : r_2 = (2\mathbf{i} + 2\mathbf{j} + 2\mathbf{k}) + \mu(2\mathbf{i} - 14\mathbf{j} + 4\mathbf{k})$

(iv) $l_1 : r_1 = (\mathbf{i} + \mathbf{j} + \mathbf{k}) + \lambda(3\mathbf{i} + 4\mathbf{j} + 5\mathbf{k})$
  $l_2 : r_2 = (4\mathbf{i} - 4\mathbf{j} + 9\mathbf{k}) + \mu(6\mathbf{i} + 8\mathbf{j} + 10\mathbf{k})$

(v) $l_1 : r_1 = (2\mathbf{i} + 2\mathbf{j} + 2\mathbf{k}) + \lambda(-\mathbf{i} + 7\mathbf{j} + 9\mathbf{k})$
  $l_2 : r_2 = (8\mathbf{i} - 8\mathbf{j} + 9\mathbf{k}) + \mu(4\mathbf{i} - 2\mathbf{j} + 2\mathbf{k})$.

Determine whether the pair of lines above are parallel, intersect or are skew. Find the position vector of the point of intersection.

Solution 26

Examine the direction ratios of the direction vectors.

(i) The direction ratios of l_1 are $2 : -3 : 1$ l_2 are $-1 : 4 : -7$ and since these direction ratios are not equal the lines are not parallel. If the lines intersect, then have a common point. Equating the coefficients of **i**, **j** and **k** we have

$$3 + 2\lambda = -2 - \mu \quad \text{or} \quad 2\lambda = -5 - \mu \quad \ldots (1)$$
$$1 - 3\lambda = 4 + 4\mu \quad \text{or} \quad 3\lambda = -3 - 4\mu \quad \ldots (2)$$
$$-4 + \lambda = 1 - 7\mu \quad \text{or} \quad \lambda = 5 - 7\mu \quad \ldots (3)$$

Solving equations (1) and (3) in order to find the values of λ and μ $2(5-7\mu) = -5-\mu$ or $10-14\mu = -5 - \mu$ or $13\mu = 15$, $\mu = \frac{15}{13}$ and substituting in (3) $\lambda = 5 - \frac{105}{13} = -\frac{40}{13}$. If the lines intersect then equation (2) is true for these values, but equation (2) is not verified and the lines are skew.

(ii) The direction ratios of l_1 and l_2 are $-1 : 2 : 3$ and $1 : -7 : 2$, the lines are not parallel.
Equating the coefficients of **i**, **j** and **k** we have

$$-2 - \lambda = 1 + \mu \quad \text{or} \quad \lambda = -3 - \mu \quad \ldots (1)$$
$$-2 + 2\lambda = 1 - 7\mu \quad \text{or} \quad 2\lambda = 3 - 7\mu \quad \ldots (2)$$
$$-4 + 3\lambda = 1 + 2\mu \quad \text{or} \quad 3\lambda = 5 + 2\mu \quad \ldots (3)$$

Solving equations (1) and (2)

$$2\lambda = -6 - 2\mu \quad \ldots (1) \times 2$$
$$2\lambda = 3 - 7\mu \quad \ldots (2)$$

$-6 - 2\mu = 3 - 7\mu$ or $5\mu = 9$ or $\mu = \frac{9}{5}$, $\lambda = -3 - \frac{9}{5} = -\frac{24}{5}$.

Substituting these values $\lambda = -\frac{24}{5}$ and $\mu = \frac{9}{5}$ in equation (3),
we find that $3\left(-\frac{24}{5}\right) \neq 5 + 2\left(\frac{9}{5}\right)$.
The lines are again skew.

(iii) The direction ratios are $2 : 2 : 6$ and $2 : -14 : 4$ or $1 : 1 : 3$ and $1 : -7 : 2$ which are not equal therefore the lines are not parallel. Equating the coefficient of **i**, **j** and **k** for the two lines we have

$$8 + 2\lambda = 2 + 2\mu \quad \text{or} \quad 2\lambda = -6 + 2\mu$$
$$\text{or} \quad \lambda = -3 + \mu \quad \ldots (1)$$
$$-8 + 2\lambda = 2 - 14\mu \quad \text{or} \quad 2\lambda = 10 - 14\mu$$
$$\text{or} \quad \lambda = 5 - 7\mu \quad \ldots (2)$$
$$18 + 6\lambda = 2 + 4\mu \quad \text{or} \quad 6\lambda = -16 + 4\mu$$
$$\text{or} \quad 3\lambda = -8 + 2\mu \ldots (3)$$

Solving (1) and (2)
$-3 + \mu = 5 - 7\mu$ or $8\mu = 8$ or $\boxed{\mu = 1}$
$\lambda = -3 + 1 = -2$ or $\boxed{\lambda = -2}$ Substituting these values in (3) $-6 = -8 + 2 = -6$, verifies these equations and the values of $\mu = 1$ and $\lambda = -2$ make the two vector equations equal $\mathbf{r}_1 = \mathbf{r}_2$, the lines therefore intersect.

$$l_1 : \mathbf{r}_1 = (8\mathbf{i} - 8\mathbf{j} + 18\mathbf{k}) - 2(2\mathbf{i} + 2\mathbf{j} + 6\mathbf{k})$$
$$= 4\mathbf{i} - 12\mathbf{j} + 6\mathbf{k}$$
$$l_2 : \mathbf{r}_2 = (2\mathbf{i} + 2\mathbf{j} + 2\mathbf{k}) + 2\mathbf{i} - 14\mathbf{j} + 4\mathbf{k}$$
$$= 4\mathbf{i} - 12\mathbf{j} + 6\mathbf{k}$$

$\mathbf{r}_1 = \mathbf{r}_2$ and the position vector of $\overrightarrow{OP} = 4\mathbf{i} - 12\mathbf{j} + 6\mathbf{k}$.

(iv) The direction ratios are for $l_1 : 3 : 4 : 5$ for $l_2 : 6 : 8 : 10$ or $3 : 4 : 5$, the direction ratios are equal and the lines are parallel.

(v) The direction ratios are for $l_1 : -1 : 7 : 9$ for $l_2 : 4 : -2 : +2$.

$$2 - \lambda = 8 + 4\mu \quad \ldots (1)$$
$$2 + 7\lambda = -8 - 2\mu \quad \ldots (2)$$
$$2 + 9\lambda = 9 + 2\mu \quad \ldots (3)$$

Solving (1) and (2) we have $\lambda = -6 - 4\mu$ from (1), $7\lambda = -10 - 2\mu$, from (2), $7(-6 - 4\mu) = -10 - 2\mu$, $-42 - 28\mu = -10 - 2\mu$, $26\mu = -32$, $\mu = -\frac{16}{13}$ and $\lambda = -6 + \frac{64}{13} = -\frac{14}{13}$. Substituting these values in (3) $2 - \frac{126}{13} \neq 9 - \frac{32}{13}$ or $-\frac{100}{13} \neq \frac{85}{13}$.
These lines are neither parallel nor do they intersect. They are skew lines. Observe that the dot product of the direction vectors is zero.

$(-\mathbf{i} + 7\mathbf{j} + 9\mathbf{k}).(4\mathbf{i} - 2\mathbf{j} + 2\mathbf{k}) = -4 - 14 + 18 = 0$
see next page. The skew lines are perpendicular.

Angle between a Pair of Line

Consider the general vector equation $\mathbf{r} = \mathbf{a} + s\mathbf{b}$ where
$\mathbf{r} = x\mathbf{i} + y\mathbf{j} + z\mathbf{k}$

$\mathbf{a} = a_1\mathbf{i} + a_2\mathbf{j} + a_3\mathbf{k}$ the position vector

$\mathbf{b} = b_1\mathbf{i} + b_2\mathbf{j} + b_3\mathbf{k}$ the direction vector

and s is the parameter.

Therefore the vector equations of any two lines are:

$l_1 : \mathbf{r} = (a_1\mathbf{i} + a_2\mathbf{j} + a_3\mathbf{k}) + \lambda\,(b_1\mathbf{i} + b_2\mathbf{j} + b_3\mathbf{k})$

$l_2 : \mathbf{r} = (a'_1\mathbf{i} + a'_2\mathbf{j} + a'_3\mathbf{k}) + \mu\,(b'_1\mathbf{i} + b'_2\mathbf{j} + b'_3\mathbf{k})$.

The angle between a pair of lines depends only on their directions and <u>not</u> on their positions.
The direction vectors are

$$\mathbf{v}_1 = b_1\mathbf{i} + b_2\mathbf{j} + b_3\mathbf{k}$$
$$\mathbf{v}_2 = b'_1\mathbf{i} + b'_2\mathbf{j} + b'_3\mathbf{k}.$$

The direction ratios are $b_1 : b_2 : b_3$ for $\mathbf{v}_1$ and $b'_1 : b'_2 : b'_3$ for $\mathbf{v}_2$.

The direction cosines are $\dfrac{b_1}{|\mathbf{v}_1|}, \dfrac{b_2}{|\mathbf{v}_1|}, \dfrac{b_3}{|\mathbf{v}_1|}$ for $\mathbf{v}_1$ and $\dfrac{b'_1}{|\mathbf{v}_2|}, \dfrac{b'_2}{|\mathbf{v}_2|}, \dfrac{b'_3}{|\mathbf{v}_2|}$ for $\mathbf{v}_2$.

Dot or Scalar Product

$\mathbf{v}_1 \cdot \mathbf{v}_2 = (b_1\mathbf{i} + b_2\mathbf{j} + b_3\mathbf{k}) \cdot (b'_1\mathbf{i} + b'_2\mathbf{j} + b'_3\mathbf{k})$
$\quad = b_1 b'_1 + b_2 b'_2 + b_3 b'_3$ where

$\boxed{\mathbf{v}_1 \cdot \mathbf{v}_2 = |\mathbf{v}_1||\mathbf{v}_2|\cos\theta}$ and $\mathbf{i}.\mathbf{j} = \mathbf{i}.\mathbf{k} = \mathbf{j}.\mathbf{k} = 0$,
$\mathbf{i}.\mathbf{i} = \mathbf{j}.\mathbf{j} = \mathbf{k}.\mathbf{k} = 1$ since $\cos 0 = 1$ and $\cos 90° = 0$

$b_1 b'_1 + b_2 b'_2 + b_3 b'_3$
$\quad = \sqrt{b_1^2 + b_2^2 + b_3^2}\,\sqrt{b'^2_1 + b'^2_2 + b'^2_3}\,\cos\theta$

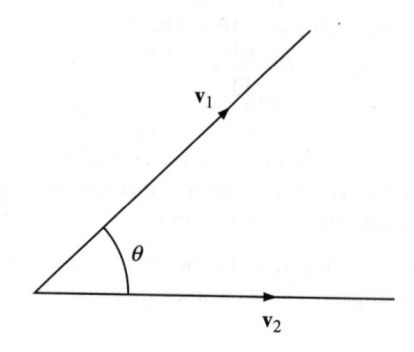

Fig. 8-I/32 Scalar product

WORKED EXAMPLE 27

Find the acute angles between l_1 and l_2, giving your answer correct to the nearest 0.1 of a degree.

(i) $l_1 : \mathbf{r} = \mathbf{i} + \mathbf{j} + \mathbf{k} + \lambda(-2\mathbf{i} + \mathbf{j} + 5\mathbf{k})$

$l_2 : \mathbf{r} = 2\mathbf{i} + 3\mathbf{j} + 4\mathbf{k} + \mu(\mathbf{i} + 2\mathbf{j} - 7\mathbf{k})$

(ii) $l_1 : \mathbf{r} = 2\mathbf{i} + 3\mathbf{j} + 5\mathbf{k} + s(\mathbf{i} + \mathbf{j} + 2\mathbf{k})$

$l_2 : \mathbf{r} = 4\mathbf{j} + 6\mathbf{k} + t(-\mathbf{i} + 2\mathbf{j} + 3\mathbf{k})$

(iii) $l_1 : \mathbf{r} = -3\mathbf{i} + 2\mathbf{j} - \mathbf{k} + \lambda(\mathbf{i} + \mathbf{j} + \mathbf{k})$

$l_2 : \mathbf{r} = 2\mathbf{i} + 3\mathbf{j} - 4\mathbf{k} + \mu(2\mathbf{i} - 3\mathbf{j} + \mathbf{k})$.

Solution 27

(i) $\mathbf{v}_1 = -2\mathbf{i} + \mathbf{j} + 5\mathbf{k}$

$\mathbf{v}_2 = \mathbf{i} + 2\mathbf{j} - 7\mathbf{k}$

$\mathbf{v}_1 \cdot \mathbf{v}_2 = (-2\mathbf{i} + \mathbf{j} + 5\mathbf{k}) \cdot (\mathbf{i} + 2\mathbf{j} - 7\mathbf{k})$
$\quad = \sqrt{(-2)^2 + (1)^2 + (5)^2}$
$\quad\quad \times \sqrt{1^2 + 2^2 + (-7)^2}\,\cos\theta$

$-2 + 2 - 35 = \sqrt{30}\,\sqrt{54}\,\cos\theta$
$\cos\theta = -0.869581991 \Rightarrow \theta = 150.4°$
$\theta = 29.6°$ the acute angle to the nearest 0.1 of a degree

(ii) $\mathbf{v}_1 = (\mathbf{i} + \mathbf{j} + 2\mathbf{k})$

$\mathbf{v}_2 = (-\mathbf{i} + 2\mathbf{j} + 3\mathbf{k})$

$\mathbf{v}_1 \cdot \mathbf{v}_2 = (\mathbf{i} + \mathbf{j} + 2\mathbf{k}) \cdot (-\mathbf{i} + 2\mathbf{j} + 3\mathbf{k})$
$\quad = \sqrt{1^2 + 1^2 + 2^2}\,\sqrt{1^2 + 2^2 + 3^2}\,\cos\theta$

$-1 + 2 + 6 = \sqrt{6}\,\sqrt{14}\,\cos\theta$

$\cos\theta = \dfrac{7}{\sqrt{6}\,\sqrt{14}}$

$\theta = 40.2°$

(iii) $\mathbf{v}_1 = \mathbf{i} + \mathbf{j} + \mathbf{k}$

$\mathbf{v}_2 = 2\mathbf{i} - 3\mathbf{j} + \mathbf{k}$

$\mathbf{v}_1 \cdot \mathbf{v}_2 = (\mathbf{i} + \mathbf{j} + \mathbf{k}) \cdot (2\mathbf{i} - 3\mathbf{j} + \mathbf{k})$
$\quad = \sqrt{1^2 + 1^2 + 1^2}\,\sqrt{2^2 + 3^2 + 1^2}\,\cos\theta$

$2 - 3 + 1 = \sqrt{3}\,\sqrt{14}\,\cos\theta$

$\cos\theta = \dfrac{0}{\sqrt{3}\,\sqrt{14}}$

$\theta = 90°$

Exercises 4

1. The line l_1 is parallel to the vector $2\mathbf{i} + 3\mathbf{j} - \mathbf{k}$ and passes through the point A with position vector $-\mathbf{i} + 3\mathbf{j} + 5\mathbf{k}$ relative to the origin O.

Write down the vector equation for l_1. The line l_2 is parallel to the vector $-\mathbf{i}+2\mathbf{j}+5\mathbf{k}$ and passes through the point B with position vector $2\mathbf{i}+3\mathbf{j}+4\mathbf{k}$ relative to the origin O.

Write down the vector equation for l_2. Determine whether the two lines l_1 and l_2 intersect or not.

2. $\mathbf{a} = 2\mathbf{i}+3\mathbf{j}+4\mathbf{k}$
   $\mathbf{b} = \mathbf{i}-2\mathbf{j}+3\mathbf{k}$
   $\mathbf{c} = 3\mathbf{i}+\mathbf{j}-2\mathbf{k}$.

   Find $\mathbf{b}+\mathbf{c}$ then find $\mathbf{a}.\mathbf{b}$, $\mathbf{a}.\mathbf{c}$ and $\mathbf{a}.(\mathbf{b}+\mathbf{c})$. State your conclusions.

3. A triangle ABC with coordinates $A(1, 2, 3)$, $B(3, 4, 6)$, $C(-1, -2, -3)$ find the vectors of $\vec{AB}, \vec{BC}$ and $\vec{AC}$. Hence calculate the area of the triangle using $\Delta = \sqrt{s(s-a)(s-b)(s-c)}$.

4. A vector makes angles of $30°, 60°$ with respect to x-axis and y-axis. Determine the angle that the vector makes with the z-axis. (You may use $\cos^2\alpha + \cos^2\beta + \cos^2\gamma = 1$).

5. The vectors $\mathbf{a}$ and $\mathbf{b}$ are given by
   $$\mathbf{a} = t\mathbf{i}+\mathbf{j}+3\mathbf{k}$$
   $$\mathbf{b} = -2\mathbf{i}+\lambda\mathbf{j}+3\mathbf{k}.$$
   (a) Find a relation between the scalars t and λ if the vectors $\mathbf{a}$ and $\mathbf{b}$ are perpendicular.
   (b) Find the values of the scalars t and λ if $\mathbf{a}$ is parallel to $\mathbf{b}$.

6. The vectors $\mathbf{a}$ and $\mathbf{b}$ are given by
   $\mathbf{a} = 7\mathbf{i}+2\lambda\mathbf{j}-9\mathbf{k}$
   $\mathbf{b} = 7\mathbf{i}+4\mathbf{j}+\mu\mathbf{k}$.
   (a) If $\mathbf{a}$ and $\mathbf{b}$ are perpendicular determine a relation between the scalars λ and μ.
   (b) If $\lambda = -1$ and $\mu = 2$, find the acute angle between the vectors $\mathbf{a}$ and $\mathbf{b}$.
   (c) What are the values of λ and μ if the vectors $\mathbf{a}$ and $\mathbf{b}$ are to be parallel.

7. With respect to an origin O, the position vectors of the points A, B and C are given as follows:
   $\mathbf{a} = \vec{OA} = 2\mathbf{i}+5\mathbf{j}-7\mathbf{k}$
   $\mathbf{b} = \vec{OB} = -3\mathbf{i}-2\mathbf{j}+4\mathbf{k}$
   $\mathbf{c} = \vec{OC} = \mathbf{i}+6\mathbf{j}+11\mathbf{k}$.
   (a) Find the vectors $\vec{AB}, \vec{AC}, \vec{BC}$.
   (b) Determine the acute angle $\angle ABC$.
   (i) using the cosine rule
   (ii) using the scalar product.

8. With respect to an origin O, the position vectors of the points P, Q and R are given as follows:
   $$\mathbf{p} = \vec{OP} = \mathbf{i}+\mathbf{j}+\mathbf{k}$$
   $$\mathbf{q} = \vec{OQ} = 2\mathbf{i}+3\mathbf{j}+4\mathbf{k}$$
   $$\mathbf{r} = \vec{OR} = -5\mathbf{i}+4\mathbf{j}-3\mathbf{k}.$$
   (a) Find the vectors $\vec{PQ}$ and $\vec{PR}$.
   (b) Calculate the acute angle $\angle RPQ$
   (i) using the cosine rule
   (ii) using the scalar product.

9. (a) The position vectors $\vec{OP}, \vec{OQ}$ and $\vec{OR}$ are given by $P(-3, 0, 4), Q(1, 3, 0), R(-3, 0, 0)$. Determine $\vec{PQ}, \vec{PR}$ and $\vec{QR}$ and hence find their moduli.
   (b) Determine the magnitudes of the following vectors and the corresponding unit vectors.
   (i) $\mathbf{a} = 3\mathbf{i}-4\mathbf{j}+5\mathbf{k}$
   (ii) $\mathbf{b} = -3\mathbf{i}+5\mathbf{k}$
   (iii) $\mathbf{r} = \mathbf{i}+\mathbf{j}+\mathbf{k}$
   (iv) $\mathbf{r} = a\mathbf{i}+b\mathbf{j}+c\mathbf{k}$
   (v) $\mathbf{p} = x\mathbf{i}+y\mathbf{j}+z\mathbf{k}$.

10. Calculate the acute angle between the vector $\mathbf{u}$ and $\mathbf{v}$ when:
    (a) $\mathbf{u} = 2\mathbf{i}-3\mathbf{j}+4\mathbf{k}$
    $\mathbf{v} = -\mathbf{i}+\mathbf{j}-\mathbf{k}$
    (b) $\mathbf{u} = (1, 2, 3)$ and $\mathbf{v} = (-1, -2, -3)$.

11. The position vectors are given by the following sets of coordinates:
    (i) $A(0, 3)$
    (ii) $B(-1, 2, 3)$
    (iii) $C(-1, -4, -9)$
    (iv) $D(0, 0, 4)$,
    (v) $E(-4, 0, 5)$.
    Write down the position vectors in the form $a\mathbf{i}+b\mathbf{j}+c\mathbf{k}$.

12. The position vectors of A, B and C are given $-7\mathbf{i}+7\mathbf{k}$, $2\mathbf{j}-3\mathbf{k}$, $-\mathbf{k}$ respectively. Write down the set of coordinates of A, B and C.

13. Find the moduli of the following vectors:
    (i) $\mathbf{u} = 2\mathbf{i} - 2\mathbf{j} + \mathbf{k}$
    (ii) $\mathbf{v} = 5\mathbf{i} + 5\mathbf{j} - 6\mathbf{k}$
    (iii) $\mathbf{w} = -\mathbf{i} - \mathbf{j} + 2\mathbf{k}$.

14. Determine the magnitude of the lines $\overrightarrow{OP}, \overrightarrow{OQ}, \overrightarrow{OR}$ where $P(-1, 4, -5)$, $Q(1, 2, 5)$, $R(-2, -4, -6)$.

15. If $\mathbf{a} = \begin{pmatrix} 4 \\ -5 \\ 6 \end{pmatrix}$, $\mathbf{b} = \begin{pmatrix} 2 \\ -3 \\ 5 \end{pmatrix}$, and $\mathbf{c} = \begin{pmatrix} 1 \\ 3 \\ 0 \end{pmatrix}$.

    Find
    (i) $|\mathbf{a}|$
    (ii) $|\mathbf{a} - \mathbf{c}|$
    (iii) $|\mathbf{b} + \mathbf{a} + \mathbf{c}|$.

16. The following vectors are given by
    (i) $\mathbf{u} = 3\mathbf{i} - \mathbf{j} - 5\mathbf{k}$
        $\mathbf{v} = -2\mathbf{i} + 5\mathbf{j} + 4\mathbf{k}$
    (ii) $\mathbf{a} = 3\mathbf{j} + \mathbf{k}$
        $\mathbf{b} = \mathbf{i} - \mathbf{k}$
    (iii) $\mathbf{w} = \mathbf{i} - \mathbf{j} - 7\mathbf{k}$
        $\mathbf{z} = -2\mathbf{i} + 2\mathbf{j} + \mathbf{k}$.

    Determine the scalar products and hence find the acute angles between the pair of vectors.

17. The magnitudes of two vectors are 1 and 2 and their acute angle is $30°$, find their scalar product.

18. Determine the following scalar products:
    (i) $\mathbf{i} \cdot \mathbf{j} \cdot \mathbf{k}$
    (ii) $\mathbf{i} \cdot \mathbf{i} = \mathbf{j} \cdot \mathbf{j} = \mathbf{k} \cdot \mathbf{k}$
    (iii) $\mathbf{i} \cdot \mathbf{i} \cdot \mathbf{i}$
    (iv) $\mathbf{i} \cdot \mathbf{j} = \mathbf{j} \cdot \mathbf{k} = \mathbf{k} \cdot \mathbf{i}$.

19. Determine $\mathbf{a} \cdot \mathbf{b}$ if
    (a) $\mathbf{a} = 2\mathbf{i} - 3\mathbf{j} - 5\mathbf{k}$, $\mathbf{b} = \mathbf{i} + 2\mathbf{j} + \mathbf{k}$
    (b) $\mathbf{a} = -\mathbf{j} - \mathbf{k}$, $\mathbf{b} = \mathbf{i} + \mathbf{j} + \mathbf{k}$
    (c) $\mathbf{a} = (3, 4, 5)$, $\mathbf{b} = (-1, -1, -1)$
    (d) $\mathbf{a} = (2, 2, 2)$, $\mathbf{b} = (2, 2, 2)$.

20. Given that $\mathbf{u} = 3t\mathbf{i} + 2t^2\mathbf{j} + \mathbf{k}$ and $\mathbf{v} = (1-t)\mathbf{i} + 3\mathbf{j} - \mathbf{k}$ where t is a scalar variable, determine
    (a) the values of t for which $\mathbf{u}$ and $\mathbf{v}$ are perpendicular,
    (b) the angle between the vector $\mathbf{u}$ and $\mathbf{v}$ when $t = -2$.

21. Given that
    $$\mathbf{a} = 2t^2\mathbf{i} + (1 - 2t)\mathbf{j} + t\mathbf{k}$$
    $$\mathbf{b} = 2t\mathbf{i} - 2t\mathbf{j} - 4t^2\mathbf{k}$$
    where t is a scalar variable, determine the value of t for which the vectors are at right angles.

22. The vectors $\mathbf{p}$ and $\mathbf{q}$ are given as $3\mathbf{i} - 5\mathbf{j} + \mathbf{k}$ and $-2\mathbf{i} - \mathbf{j} + 2\mathbf{k}$ determine the scalar product $\mathbf{p} \cdot \mathbf{q}$ and hence find the acute angle between the vectors.

23. The position vectors of the points A and B are given
    $$\mathbf{a} = \mathbf{i} + 3\mathbf{j} + \mathbf{k}$$
    $$\mathbf{b} = 2\mathbf{i} - 2\mathbf{j} - 3\mathbf{k}$$
    determine the scalar product $\mathbf{a} \cdot \mathbf{b}$ and hence find the tangent of the acute angle between the vectors.

24. The following position vectors of the points A, B, C, D and E are given:
    $$\mathbf{a} = 3\mathbf{j} - 5\mathbf{k}$$
    $$\mathbf{b} = \mathbf{i} + 2\mathbf{j} - 3\mathbf{k}$$
    $$\mathbf{c} = -2\mathbf{i} - \mathbf{j} + \mathbf{k}$$
    $$\mathbf{d} = 3\mathbf{i} + 2\mathbf{j} + \mathbf{k}$$
    $$\mathbf{e} = 3\mathbf{i} + 4\mathbf{j} + 5\mathbf{k}.$$

    Determine:
    (a) (i) $2\mathbf{a} - 3\mathbf{e}$
        (ii) $\mathbf{b} + 3\mathbf{d} + \mathbf{e}$
        (iii) $2\mathbf{b} - 4\mathbf{d} + 2\mathbf{e}$.
    (b) (i) $|\mathbf{b} \cdot \mathbf{e}|$
        (ii) $|\mathbf{a} - \mathbf{e}|$
        (iii) $|\mathbf{d} - \mathbf{a}|$
        (iv) $|\mathbf{d} \cdot \mathbf{e}|$
        (v) $|(\mathbf{a} \cdot \mathbf{b}) \cdot \mathbf{c}|$

(c) the direction cosines of each vector.

(d) the angle

   (i) between **a** and **e**

   (ii) between 2**a** and **d**

   (iii) between **c** and 2**e**.

25. The vector equations of the three lines are given below:

$$l_1 : \mathbf{r} = 2\mathbf{i} + 3\mathbf{j} + 4\mathbf{k} + \lambda(2\mathbf{i} - 3\mathbf{j} + 5\mathbf{k})$$

$$l_2 : \mathbf{r} = 2\mathbf{i} - 5\mathbf{j} + \mathbf{k} + \mu(4\mathbf{i} - 6\mathbf{j} + 10\mathbf{k})$$

$$l_3 : \mathbf{r} = -\mathbf{i} + \mathbf{j} + 3\mathbf{k} + \nu(\mathbf{i} - \mathbf{j} + 2\mathbf{k}).$$

Find which pair of lines

   (i) are parallel to each other

   (ii) are skew

   (iii) intersect with each other.

26. Find the perpendicular distance from the point $A(2, -3, 4)$ to the line with vector equation $\mathbf{r} = -\mathbf{i} + 2\mathbf{j} + 4\mathbf{k} + \lambda(2\mathbf{i} - 3\mathbf{j} + 5\mathbf{k})$.

27. Find the perpendicular distance from the point $A(1, 2, 3)$ to the line with vector equation $\mathbf{r} = \mathbf{i} + 3\mathbf{j} + 5\mathbf{k} + \mu(4\mathbf{i} + 3\mathbf{j} + 2\mathbf{k})$.

28. A line passes through the point (x_1, y_1, z_1) and is parallel to the direction vector $a\mathbf{i} + b\mathbf{j} + c\mathbf{k}$. Write down the vector equation of the line in parametric form and hence find the cartesian form of the line.

29. Prove that a line that passes through a point $A(x_1, y_1, z_1)$ and is parallel to the vector $\mathbf{b} = a\mathbf{i} + b\mathbf{j} + c\mathbf{k}$ is given by the vector equation

$$\mathbf{r} = \mathbf{a} + \lambda \mathbf{b}$$

where λ is a parameter.

30. A line passes through two points $A(x_1, y_1, z_1)$ and $B(x_2, y_2, z_2)$ show that the vector equation of the line is given

$$\mathbf{r} = (x_1\mathbf{i} + y_1\mathbf{j} + z_1\mathbf{k})$$
$$+ \lambda[(x_2 - x_1)\mathbf{i} + (y_2 - y_1)\mathbf{j} + (z_2 - z_1)\mathbf{k}].$$

31. Write down the vector equations of the lines corresponding to the following cartesian equations:

   (i) $\dfrac{x+1}{1} = \dfrac{y-2}{2} = \dfrac{z+3}{3}$

   (ii) $\dfrac{x-3}{-3} = \dfrac{y+1}{5} = \dfrac{z-5}{7}$

   (iii) $\dfrac{x-2}{-1} = \dfrac{y-1}{-2} = \dfrac{z-4}{-3}$

   (iv) $\dfrac{x-1}{0} = \dfrac{y-2}{0} = \dfrac{z}{2}.$

32. The vector equations of two lines l_1, l_2 are given:

$$l_1 : \mathbf{r} = (2\mathbf{i} - 3\mathbf{j} + 4\mathbf{k}) + \lambda(-\mathbf{i} - 3\mathbf{j} + 2\mathbf{k})$$

$$l_2 : \mathbf{r} = (3\mathbf{i} + 2\mathbf{k}) + \mu(\mathbf{i} - 2\mathbf{j} - 3\mathbf{k}).$$

Show that the lines intersect and find the position vector of the point of intersection.

33. The cartesian equations of two lines are given $y = m_1x + c_1$, $y = m_2x + c_2$ find the vector equations of the lines and hence show that the acute angle of the intersection of the lines is given $\tan^{-1}\dfrac{m_1 - m_2}{1 + m_1m_2}$.

34. Determine by means of vectors the perpendicular distance from a point $A(x_1, y_1)$ to the line $ax + by + c = 0$.

35. Find the distance between the pair of parallel lines

$$\mathbf{r} = (\mathbf{i} + \mathbf{j} + \mathbf{k}) + \lambda(-2\mathbf{i} - 3\mathbf{j} - 4\mathbf{k}) : l_1$$

$$\mathbf{r} = (2\mathbf{i} - 3\mathbf{j} - \mathbf{k}) + \mu(-2\mathbf{i} - 3\mathbf{j} - 4\mathbf{k}) : l_2$$

36. Find the distance between the pair of parallel lines

$$\dfrac{x-1}{2} = \dfrac{y+2}{3} = \dfrac{z-3}{1}$$

$$\dfrac{x+3}{2} = \dfrac{y-3}{3} = \dfrac{z+1}{1}.$$

5

Coordinate Geometry in 3 Dimensions

The Equation of a Plane

A plane may be located if we know a fixed point on the plane and the plane is known to be perpendicular to a given direction.

Let $A(-2, 3, 5)$ be a fixed point on the plane which is perpendicular to the line with direction ratios $3 : 4 : 7$.

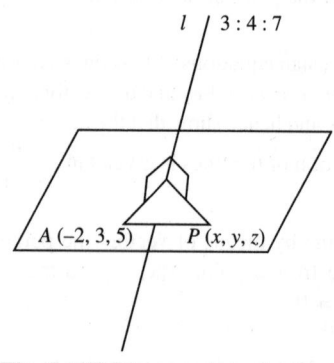

Fig. 8-I/33 The equation of a plane.

Since the line l is perpendicular to the plane then it is perpendicular to any line in the plane.

Let $P(x, y, z)$ be a general point on the plane, then AP is perpendicular to the line l.

The direction ratios of AP are $(x + 2) : (y - 3) : (z - 5)$ and since the line is perpendicular to AP

$$3(x + 2) + 4(y - 3) + 7(z - 5) = 0$$
$$3x + 6 + 4y - 12 + 7z - 35 = 0$$
$$\boxed{3x + 4y + 7z = 41}$$

therefore $P(x, y, z)$ is a point on the plane
$$3x + 4y + 7z = 41$$

In general, $\boxed{ax + by + cz = d}$ represents the cartesian equation of a plane, where $a : b : c$ are the direction ratios of a normal to the plane.

If a plane is normal to a line with direction ratios $a : b : c$ and contains the point (x_1, y_1, z_1) then the equation of the plane is written directly as

$$ax + by + cz = ax_1 + by_1 + cz_1.$$

WORKED EXAMPLE 28

Find the equation of the plane for the point $A(0, -3, -7)$ on the plane which is perpendicular to the line with direction ratios $1 : 2 : 3$.

Solution 28

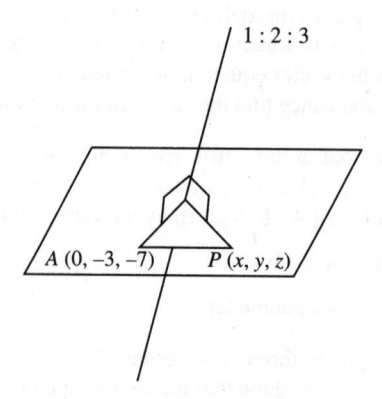

Fig. 8-I/34 Equation of the plane

The direction ratios of AP are $x : (y + 3) : (z + 7)$ and since the line is perpendicular to AP

$$1(x) + 2(y + 3) + 3(z + 7) = 0$$
$$x + 2y + 6 + 3z + 21 = 0$$

therefore this is $\boxed{x + 2y + 3z = -27}$

the equation of the plane and $P(x, y, z)$ is a point on it.

27

WORKED EXAMPLE 29

Find the equation of the plane which contains three points with coordinates $A(1, 2, 3)$, $B(-2, 3, -1)$, $C(-1, 4, 2)$.

Solution 29

The equation of the plane is $ax + by + cz = d$, where $a : b : c$ are the direction ratios of the line which is perpendicular to the plane.

Since $A(1, 2, 3)$ lies on the plane then this will satisfy the equation.

$$a + 2b + 3c = d \quad \ldots(1)$$

Since $B(-2, 3, -1)$ lies also on the plane then this will also satisfy the equation

$$-2a + 3b - c = d \quad \ldots(2)$$

and finally $C(-1, 4, 2)$ lies on the plane then

$$-a + 4b + 2c = d \quad \ldots(3)$$

Solving equations (1), (2) and (3) in terms of d

$$a + 2b + 3c = d \quad \ldots(1)$$
$$-2a + 3b - c = d \quad \ldots(2)$$
$$-a + 4b + 2c = d \quad \ldots(3)$$

adding (1) and (3)

$$6b + 5c = 2d \quad \ldots(4)$$

$$\begin{array}{ll}(1) \times 2 & 2a + 4b + 6c = 2d \\ (2) & -2a + 3b - c = d \\ \hline & 7b + 5c = 3d \end{array} \quad \ldots(5)$$

(5) − (4) $\boxed{b = d}$

Substituting this value in (4)

$$6d + 5c = 2d$$

$$5c = -4d \quad \boxed{c = \frac{-4}{5}d}$$

From (1) $a + 2b + 3c = d$

$$a + 2d - \frac{12}{5}d = d \qquad a = d - 2d + \frac{12}{5}d$$

$$a = \frac{5d - 10d + 12d}{5} \qquad \boxed{a = \frac{7}{5}d}$$

$$\therefore a = \frac{7}{5}d, \; b = d, \; c = -\frac{4}{5}d$$

The plane through A, B, and C is given by

$$\frac{7}{5}dx + dy - \frac{4}{5}dz = d$$

$$\boxed{7x + 5y - 4z = 5}$$

If a plane contains the origin, then $O(0, 0, 0)$ satisfies the equation $ax + by + cz = d$ where $d = 0$.

Any plane through the origin has an equation

$$\boxed{ax + by + cz = 0}$$

WORKED EXAMPLE 30

Find the equation of the plane which contains the two lines

$$l_1 : \frac{x+2}{4} = \frac{y+6}{3} = \frac{z-2}{2}$$

$$l_2 : \frac{x-3}{1} = \frac{y+1}{2} = \frac{z-7}{3}.$$

Solution 30

The lines must be parallel or intersecting.

Let $l_1 : \dfrac{x+2}{4} = \dfrac{y+6}{3} = \dfrac{z-2}{2} = t \quad \ldots(1)$

and $l_2 : \dfrac{x-3}{1} = \dfrac{y+1}{2} = \dfrac{z-7}{3} = s \quad \ldots(2)$

where t and s are scalar parameters.

From (1) $3(x + 2) = 4(y + 6)$

From (2) $2(x - 3) = y + 1$

$3x - 4y = 24 - 6 \qquad 2x - y = 6 + 1$

$\boxed{3x - 4y = 18} \quad \ldots(3) \qquad \boxed{2x - y = 7} \quad \ldots(4)$

Solve (3) and (4)

$3x - 4y = 18 \qquad \ldots(3)$

$\underline{-8x + 4y = -28} \qquad \ldots(4) \times (-4)$

$-5x = -10$

$\boxed{x = 2}$ and $\boxed{y = -3}$

From (1) and (2)

$2(y + 6) = 3(z - 2) \qquad 3(y + 1) = 2(z - 7)$

$2y - 3z = -12 - 6 \qquad 3y - 2z = -3 - 14$

$\boxed{2y - 3z = -18} \quad \ldots(5) \qquad \boxed{3y - 2z = -17} \quad \ldots(6)$

$(5) \times 3 \quad 6y - 9z = -54$ ...(7)

$(6) \times (-2) \quad -6y + 4z = 34$ ...(8)

adding (7) and (8) $\quad -5z = -20,$ $\boxed{z = 4}$ $\boxed{y = -3}$

Equation (1) and (2) are consistent

$\boxed{x = 2}$ $\boxed{y = -3}$ and $\boxed{z = 4}$

therefore the lines intersect at $P(2, -3, 4)$

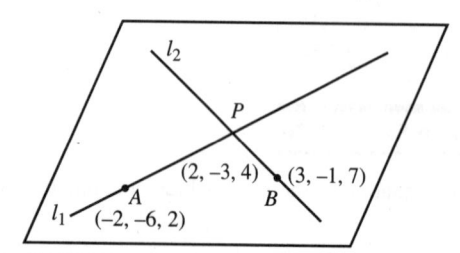

Fig. 8-I/35 Equation of a plane containing two lines

$l_1 : \dfrac{x+2}{4} = \dfrac{y+6}{3} = \dfrac{z-2}{2} = t = 1$

$l_2 : \dfrac{x-3}{1} = \dfrac{y+1}{2} = \dfrac{z-7}{3} = s = -1$

when $t = 0, x = -2, y = -6, z = 2; A(-2, -6, 2)$

when $s = 0, x = 3, y = -1, z = 7; B(3, -1, 7)$.

For each value of t corresponds to one and only one point on l_1. For each value of s corresponds to one and only one point on l_2. Find the plane containing the points $A(-2, -6, 2), B(3, -1, 7), C(2, -3, 4)$.

Since these points lie on the plane, then will satisfy the equation of the plane

$ax + by + cz = d$

$-2a - 6b + 2c = d$ ...(1)

$3a - b + 7c = d$ ...(2)

$2a - 3b + 4c = d$ ...(3)

Adding (1) and (3)

$\boxed{-9b + 6c = 2d}$ ...(4)

adding (1) $\times 3$ and (2) $\times 2$

$-6a - 18b + 6c = 3d$
$\underline{6a - 2b + 14c = 2d}$
$-20b + 20c = 5d$

or $\boxed{-4b + 4c = d}$ ...(5)

$(4) \times 2 \quad -18b + 12c = 4d$

$(5) \times -3 \quad \underline{+12b - 12c = -3d}$

$\qquad \qquad -6b = d$

$\boxed{b = -\dfrac{d}{6}}$ ...(6)

$4c = d + 4b = d - \dfrac{4d}{6} = \dfrac{2d}{6} = \dfrac{d}{3}$

$\boxed{c = \dfrac{d}{12}}$ ...(7)

From (1) $-2a - 6\left(-\dfrac{d}{6}\right) + 2\left(\dfrac{d}{12}\right) = d$

$-2a = d - d - \dfrac{1}{6}d$

$\boxed{a = \dfrac{1}{12}d}$ ...(8)

Substituting (6), (7) and (8) in plane equation

$\dfrac{1}{12}dx - \dfrac{1}{6}dy + \dfrac{1}{12}dz = d$

therefore the equation of the plane is

$\boxed{x - 2y + z = 12}$

This example can be solved alternatively. Let l_3 be a line perpendicular to both lines l_1 and l_2, with direction ratio $a : b : c$.

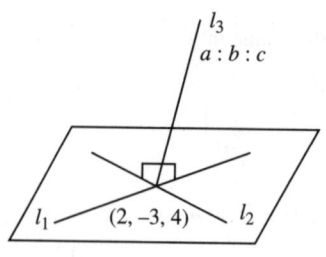

$l_1 : \dfrac{x+2}{4} = \dfrac{y+6}{3} = \dfrac{z-2}{2}$

$l_2 : \dfrac{x-3}{1} = \dfrac{y+1}{2} = \dfrac{z-7}{3}$

Fig. 8-I/36 Equation of a plane containing two lines.

Since l_3 is perpendicular to l_1 then

$4a + 3b + 2c = 0$ ...(1)

Since l_3 is perpendicular to l_2 then

$a + 2b + 3c = 0$ ...(2)

Solving (1) and (2) simultaneously

(1) $\qquad 4a + 3b + 2c = 0$

(2) × −4 $\quad \dfrac{-4a - 8b - 12c = 0}{-5b - 10c = 0}$

$$\dfrac{b}{c} = -\dfrac{2}{1}$$

From (2) $\quad a - 4c + 3c = 0, a = c$

$$\dfrac{a}{c} = \dfrac{1}{1}$$

$$\dfrac{\tfrac{a}{c}}{\tfrac{b}{c}} = \dfrac{1}{-2} = -\dfrac{1}{2}$$

$$\dfrac{a}{b} = -\dfrac{1}{2}$$

$a : b : c = 1 : -2 : 1$.

The intersection of l_1 and l_2 is $(2, -3, 4)$

$x - 2y + z = (1)(2) + (-2)(-3) + (1)(4) = 12$

$$\boxed{x - 2y + z = 12}$$

The Perpendicular Distance of the Point $A(x_1, y_1, z_1)$ from the Plane $ax + by + cz = d$ is

$$\left| \dfrac{ax_1 + by_1 + cz_1 - d}{\sqrt{a^2 + b^2 + c^2}} \right|.$$

Let the equation of a plane Π be $ax + by + cz = d$ and the point $A(x_1, y_1, z_1)$.

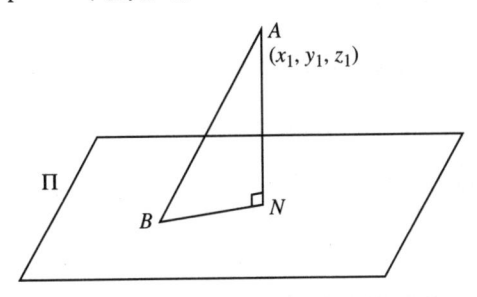

Fig. 8-I/37 The equation of a plane.

AN is drawn perpendicular to the plane Π, B is a point on the plane, so that BN is perpendicular to AN.

The direction ratios of AN are $a : b : c$ and the corresponding direction cosines are $l = \dfrac{a}{\sqrt{a^2 + b^2 + c^2}}$,

$m = \dfrac{b}{\sqrt{a^2 + b^2 + c^2}}$, $n = \dfrac{c}{\sqrt{a^2 + b^2 + c^2}}$.

Equation of a Plane

$\hat{\mathbf{n}}$ = unit vector perpendicular to the plane

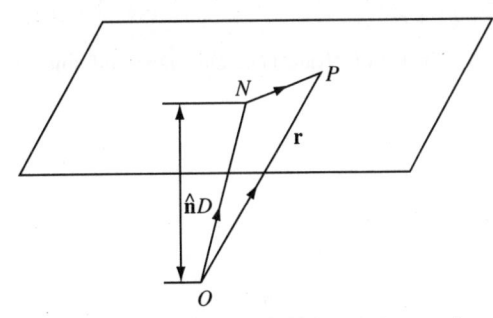

Fig. 8-I/38 Equation of the plane in the form $\mathbf{r} \cdot \hat{\mathbf{n}} = D$.

Draw a perpendicular from the origin O to the plane, ON, where N is the foot $\overrightarrow{ON} = \hat{\mathbf{n}}D$. Take any point on the plane P, where $\overrightarrow{NP}$ is perpendicular to $\overrightarrow{ON}$. The scalar product or the dot product is given $\overrightarrow{NP} \cdot \overrightarrow{ON} = 0$.

If $\mathbf{r}$ is the position vector of P, $\overrightarrow{NP} = \mathbf{r} - \hat{\mathbf{n}}D$

$(\mathbf{r} - \hat{\mathbf{n}}D) \cdot \hat{\mathbf{n}}D = 0$

$\mathbf{r} \cdot \hat{\mathbf{n}} - \hat{\mathbf{n}} \cdot \hat{\mathbf{n}}D = 0$

but $\qquad \hat{\mathbf{n}} \cdot \hat{\mathbf{n}} = 1$

$\boxed{\mathbf{r} \cdot \hat{\mathbf{n}} = D} \Rightarrow \boxed{\mathbf{r} \cdot \hat{\mathbf{n}} = d}$ where $\dfrac{D}{|\mathbf{n}|} = d$

the standard form of the vector equation of the plane.

Distance of a Plane from the Origin

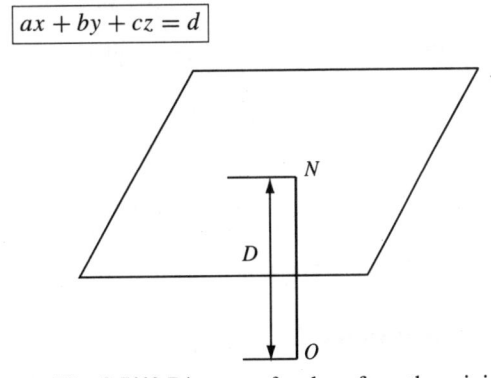

$\boxed{ax + by + cz = d}$

Fig. 8-I/40 Distance of a plane from the origin.

ON is perpendicular to the plane Π, the direction cosines $l : m : n$ where

$$l = \frac{a}{\sqrt{a^2+b^2+c^2}}, \quad m = \frac{b}{\sqrt{a^2+b^2+c^2}} \text{ and } n = \frac{c}{\sqrt{a^2+b^2+c^2}}.$$

The coordinates of N are (Dl, Dm, Dn) and since N lies on Π,

$$D = \frac{d}{\sqrt{a^2+b^2+c^2}}.$$

Dividing each term, $ax + by + cz = d$, by $\sqrt{a^2+b^2+c^2}$, we have

$$\frac{a}{\sqrt{a^2+b^2+c^2}}x + \frac{b}{\sqrt{a^2+b^2+c^2}}y + \frac{c}{\sqrt{a^2+b^2+c^2}}z = D$$

$$\boxed{lx + my + nz = D}$$

The Distance between Two Parallel Planes

$$ax + by + cz = d_2$$
$$ax + by + cz = d_1$$

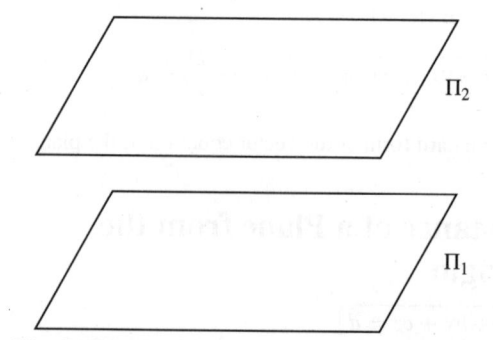

Fig. 8-I/39 Distance between two parallel planes.

The distance from the origin to the plane Π_2 is $\dfrac{d_2}{\sqrt{a^2+b^2+c^2}}$ and the distance from the origin to the plane Π_1 is $\dfrac{d_1}{\sqrt{a^2+b^2+c^2}}$. Therefore the distance between the two parallel planes is $\dfrac{d_2 - d_1}{\sqrt{a^2+b^2+c^2}}$.

WORKED EXAMPLE 31

Find the distance between the two parallel planes given by the cartesian equation $x + 3y - \sqrt{15}z = 6$, and $x + 3y - \sqrt{15}z = 1$.

Solution 31

$$\frac{d_2 - d_1}{\sqrt{a^2+b^2+c^2}} = \frac{6 - 1}{\sqrt{1^2+3^2+15}} = \frac{5}{5} = 1.$$

The Parametric Form of the Vector Equation of a Plane

Consider three points A, B and C with position vectors **a**, **b** and **c** respectively.

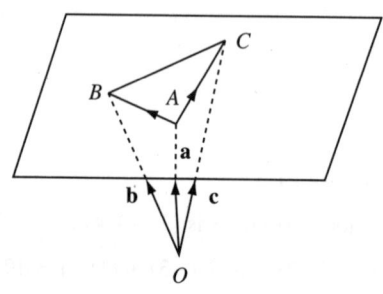

Fig. 8-I/41 Parametric form of a plane.

$\overrightarrow{AB} = \mathbf{b} - \mathbf{a}, \overrightarrow{AC} = \mathbf{c} - \mathbf{a}$.

The plane, through the point with position vector **a** and parallel to $\overrightarrow{AB}$ and $\overrightarrow{AC}$, has a vector equation

$$\boxed{\mathbf{r} = \mathbf{a} + s\overrightarrow{AB} + t\overrightarrow{AC}}$$

where s and t are independent parameters

$$\mathbf{r} = \mathbf{a} + s(\mathbf{b} - \mathbf{a}) + t(\mathbf{c} - \mathbf{a})$$

$$\mathbf{r} = \mathbf{a}(1 - s - t) + s\mathbf{b} + t\mathbf{c}$$

$$\boxed{\mathbf{r} = \lambda\mathbf{a} + s\mathbf{b} + t\mathbf{c}} \qquad \ldots(1)$$

where $\lambda = 1 - s - t$, therefore $\lambda + s + t = 1$.

Equation (1) represents the parametric form of the vector equation of a plane through three points with position vectors **a**, **b**, and **c**.

WORKED EXAMPLE 32

Find the vector equation of the plane containing the points $A(2, 1, 0)$, $B(3, -1, 1)$, and $C(0, -2, -1)$,

(a) in parametric form

(b) in cartesian form,

(c) in scalar product form.

Solution 32

(a) The parametric equation of this plane is

$$\mathbf{r} = \lambda(2\mathbf{i} + \mathbf{j}) + \mu(3\mathbf{i} - \mathbf{j} + \mathbf{k}) + \nu(-2\mathbf{j} - \mathbf{k})$$

where $\lambda + \mu + \nu = 1$.

$\nu = 1 - \lambda - \mu$

$\mathbf{r} = \lambda(2\mathbf{i}+\mathbf{j})+\mu(3\mathbf{i}-\mathbf{j}+\mathbf{k})+(1-\lambda-\mu)(-2\mathbf{j}-\mathbf{k})$

$\mathbf{r} = (2\lambda + 3\mu)\mathbf{i} + (\lambda - \mu + 2\lambda + 2\mu - 2)\mathbf{j}$
$\qquad + (\mu - 1 + \lambda + \mu)\mathbf{k}$

$$\boxed{\mathbf{r} = (2\lambda + 3\mu)\mathbf{i} + (3\lambda + \mu - 2)\mathbf{j} + (2\mu + \lambda - 1)\mathbf{k}}$$

(b) $x = 2\lambda + 3\mu$

$y = 3\lambda + \mu - 2$

$z = 2\mu + \lambda - 1$.

Eliminating λ and μ from these equations

$x = 2\lambda + 3\mu, \; -3y = -9\lambda - 3\mu + 6$

$$\boxed{x - 3y = -7\lambda + 6}$$

$-2y = -6\lambda - 2\mu + 4, \qquad z = 2\mu + \lambda - 1$

$$\boxed{-2y + z = -5\lambda + 3}$$

$x - 3y = -\dfrac{7(-2y + z - 3)}{-5} + 6$

$+5x - 15y = 7(-2y + z - 3) + 30$

$5x - 15y + 14y - 7z + 21 - 30 = 0$

$$\boxed{5x - y - 7z = 9}$$

(c) the cartesian form of the plane. Therefore the scalar form of the equation is

$$\boxed{\mathbf{r} \cdot (5\mathbf{i} - \mathbf{j} - 7\mathbf{k}) = 9}.$$

Plane Passing through a Given Point and Perpendicular to a Given Direction

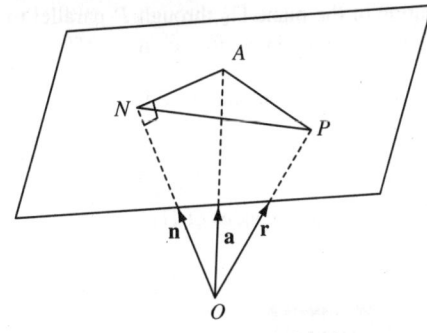

Fig. 8-I/42 Equation of a plane.

Let A be the given point and its position vector is $\overrightarrow{OA} = \mathbf{a}$.

Let the unit vector in the given direction be $\hat{\mathbf{n}}$, $\overrightarrow{ON} = n\hat{\mathbf{n}}$ where n is the length of $\overrightarrow{ON}$. As $\overrightarrow{ON}$ is perpendicular to the plane it is perpendicular to AP.

$(\mathbf{r} - \mathbf{a}) \cdot \hat{\mathbf{n}} = 0$

$$\boxed{\mathbf{r} \cdot \hat{\mathbf{n}} = \mathbf{a} \cdot \hat{\mathbf{n}}}$$

the vector equation of the plane

$$\boxed{\mathbf{r} \cdot \hat{\mathbf{n}} = n}$$

the perpendicular form, where n is positive as the scalar product $\mathbf{a} \cdot \hat{\mathbf{n}}$ is positive and the angle between the vectors is acute.

$ON = n$, the projection of OA.

Perpendicular Distance of a Point from a Plane

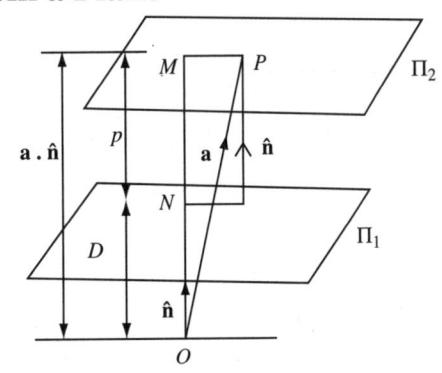

Fig. 8-I/43 Perpendicular distance of a point from a plane

Consider a plane Π_1 with equation $\mathbf{r} \cdot \hat{\mathbf{n}} = D$ where D is the distance from the origin to the plane Π_1, and let a point P with position vector $(\overrightarrow{OP})$.

The equation of the plane Π_2 through P parallel to the plane Π_1 has an equation $\mathbf{r} \cdot \hat{\mathbf{n}} = \mathbf{a} \cdot \hat{\mathbf{n}} = OM$.

MN = distance of P from the plane Π

$$p = \mathbf{a} \cdot \hat{\mathbf{n}} - D$$

P and O are on opposite sides of Π_1.

WORKED EXAMPLE 33

Find the distance of the point with position vector $\mathbf{i}+\mathbf{j}+\mathbf{k}$ from a plane Π with equation $\mathbf{r} \cdot (2\mathbf{i} + 3\mathbf{j} + 4\mathbf{k}) = 5$.

Solution 33

$p = \mathbf{a} \cdot \hat{\mathbf{n}} - D$

$= (\mathbf{i} + \mathbf{j} + \mathbf{k}) \cdot \dfrac{(2\mathbf{i} + 3\mathbf{j} + 4\mathbf{k})}{\sqrt{2^2 + 3^2 + 4^2}} - \dfrac{5}{\sqrt{2^2 + 3^2 + 4^2}}$

$= (\mathbf{i} + \mathbf{j} + \mathbf{k}) \cdot \dfrac{(2\mathbf{i} + 3\mathbf{j} + 4\mathbf{k})}{\sqrt{29}} - \dfrac{5}{\sqrt{29}}$

$= \dfrac{1}{\sqrt{29}}(2 + 3 + 4) - \dfrac{5}{\sqrt{29}}$

$= \dfrac{4}{\sqrt{29}}$

therefore the point P and the origin are on opposite sides of the plane.

WORKED EXAMPLE 34

Find the distance of the point (1, 2, 3) from the plane (a) and find the distances of the point (1, 1, 1) from the planes (b) and (c).

(a) $\mathbf{r} \cdot (-\mathbf{i} - \mathbf{j} - \mathbf{k}) = 2$

(b) $\mathbf{r} \cdot (2\mathbf{i} - 3\mathbf{j} + 5\mathbf{k}) = 10$

(c) $\mathbf{r} \cdot (3\mathbf{i} + 4 + 7\mathbf{k}) = 5$.

Solution 34

(a) $p = \mathbf{a} \cdot \hat{\mathbf{n}} - D$

$= (\mathbf{i} + 2\mathbf{j} + 3\mathbf{k}) \cdot \dfrac{(-\mathbf{i} - \mathbf{j} - \mathbf{k})}{\sqrt{3}} - \dfrac{2}{\sqrt{3}}$

$= -\dfrac{1}{\sqrt{3}} - \dfrac{2}{\sqrt{3}} - \dfrac{3}{\sqrt{3}} - \dfrac{2}{\sqrt{3}} = -\dfrac{8}{\sqrt{3}}$

the negative sign indicates that the point and the origin are on the same side of the plane.

(b) $p = \mathbf{a} \cdot \hat{\mathbf{n}} - D$

$= (\mathbf{i} + \mathbf{j} + \mathbf{k}) \cdot \dfrac{(2\mathbf{i} - 3\mathbf{j} + 5\mathbf{k})}{\sqrt{38}} - \dfrac{10}{\sqrt{38}}$

$= \dfrac{2}{\sqrt{38}} - \dfrac{3}{\sqrt{38}} + \dfrac{5}{\sqrt{38}} - \dfrac{10}{\sqrt{38}} = -\dfrac{6}{\sqrt{38}}$

the negative sign indicates that the point and the origin are on the same side of the plane.

(c) $p = \mathbf{a} \cdot \hat{\mathbf{n}} - D$

$= (\mathbf{i} + \mathbf{j} + \mathbf{k}) \cdot \dfrac{(3\mathbf{i} + 4\mathbf{j} + 7\mathbf{k})}{\sqrt{74}} - \dfrac{5}{\sqrt{74}}$

$= \dfrac{3}{\sqrt{74}} + \dfrac{4}{\sqrt{74}} + \dfrac{7}{\sqrt{74}} - \dfrac{5}{\sqrt{74}} = \dfrac{9}{\sqrt{74}}$

the positive sign indicates that the point and the origin are on opposite sides of the plane.

WORKED EXAMPLE 35

(a) Find the perpendicular distance of the point $(1, -1, -1)$ from the plane $\mathbf{r} \cdot (\mathbf{i} + \mathbf{j} + \mathbf{k}) = 1$

(b) Find the perpendicular distance of the point $(-1, 2, -3)$ from the plane $\mathbf{r} \cdot (\mathbf{i} + \mathbf{j} - \mathbf{k}) = 2$.

Solution 35

(a) $p = \mathbf{a} \cdot \hat{\mathbf{n}} - D$

$= (\mathbf{i} - \mathbf{j} - \mathbf{k}) \cdot \dfrac{(\mathbf{i} + \mathbf{j} + \mathbf{k})}{\sqrt{3}} - \dfrac{1}{\sqrt{3}}$

$= \dfrac{1}{\sqrt{3}} - \dfrac{1}{\sqrt{3}} - \dfrac{1}{\sqrt{3}} - \dfrac{1}{\sqrt{3}} = -\dfrac{2}{\sqrt{3}}$

the point and the origin are on the same side.

(b) $p = \mathbf{a} \cdot \hat{\mathbf{n}} - D$

$= (-\mathbf{i} + 2\mathbf{j} - 3\mathbf{k}) \cdot \dfrac{(\mathbf{i} + \mathbf{j} - \mathbf{k})}{\sqrt{3}} - \dfrac{2}{\sqrt{3}}$

$= -\dfrac{1}{\sqrt{3}} + \dfrac{2}{\sqrt{3}} + \dfrac{3}{\sqrt{3}} - \dfrac{2}{\sqrt{3}} = \dfrac{2}{\sqrt{3}}$

the point and the origin are on opposite sides of the plane.

Worked Example 36

Determine whether the points $(1, -2, 1)$, $(-2, 1, 3)$ are on the same or opposite sides of the plane
$\mathbf{r} \cdot (\mathbf{i} + 2\mathbf{j} - \mathbf{k}) = 1$.

Solution 36

$p = \mathbf{a} \cdot \hat{\mathbf{n}} - D$

$= (\mathbf{i} - 2\mathbf{j} + \mathbf{k}) \cdot \dfrac{(\mathbf{i} + 2\mathbf{j} - \mathbf{k})}{\sqrt{6}} - \dfrac{1}{\sqrt{6}}$

$= \dfrac{1}{\sqrt{6}} - \dfrac{4}{\sqrt{6}} - \dfrac{1}{\sqrt{6}} - \dfrac{1}{\sqrt{6}} = -\dfrac{5}{\sqrt{6}}$

$p = \mathbf{a} \cdot \hat{\mathbf{n}} - D$

$= (-2\mathbf{i} + \mathbf{j} + 3\mathbf{k}) \cdot \dfrac{(\mathbf{i} + 2\mathbf{j} - \mathbf{k})}{\sqrt{6}} - \dfrac{1}{\sqrt{6}}$

$= -\dfrac{2}{\sqrt{6}} + \dfrac{2}{\sqrt{6}} - \dfrac{3}{\sqrt{6}} - \dfrac{1}{\sqrt{6}} = -\dfrac{4}{\sqrt{6}}$

the points are on the same side of the plane since they are both of the same sign and the points and the origin are on the same side.

The Angle between Two Planes

Consider two planes Π_1 and Π_2 whose vector equations are $\mathbf{r} \cdot \hat{\mathbf{n}}_1 = D_1$ and $\mathbf{r} \cdot \hat{\mathbf{n}}_2 = D_2$.

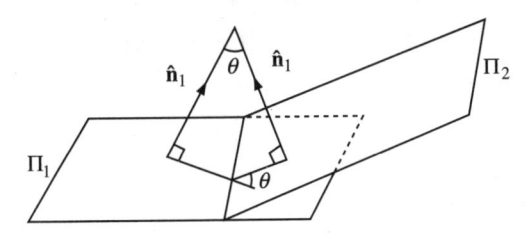

Fig. 8-I/44 Angle between two planes

The angle between the planes Π_1 and Π_2 is equal to the angle between the normals to Π_1 and Π_2, θ

$$\boxed{\cos \theta = \hat{\mathbf{n}}_1 . \hat{\mathbf{n}}_2}$$

Worked Example 37

Find the angle, to the nearest degree, between the planes whose vector equations are $\mathbf{r} \cdot (2\mathbf{i} + 3\mathbf{j} + 4\mathbf{k}) = 5$ and $\mathbf{r} \cdot (+\mathbf{i} + 2\mathbf{j} + 3\mathbf{k}) = 7$.

Solution 37

$\cos \theta = \dfrac{(2\mathbf{i} + 3\mathbf{j} + 4\mathbf{k})}{\sqrt{2^2 + 3^2 + 4^2}} \cdot \dfrac{(+\mathbf{i} + 2\mathbf{j} + 3\mathbf{k})}{\sqrt{1^2 + 2^2 + 3^2}}$

$= \dfrac{+2 + 6 + 12}{\sqrt{29}\sqrt{14}}$

$= \dfrac{20}{\sqrt{29}\sqrt{14}}$

$\theta = 7°$ to the nearest integer.

Worked Example 38

Determine the condition that two planes (i) are parallel (ii) are perpendicular.

Solution 38

(i) For the planes to be parallel the angle must be zero, $\cos \theta = \cos 0° = 1 = \hat{\mathbf{n}}_1 . \hat{\mathbf{n}}_2$, the unit vectors must be equal $\boxed{\hat{\mathbf{n}}_1 = \hat{\mathbf{n}}_2}$

(ii) For the planes to be perpendicular the angle must be $90°$, $\cos 90° = 0$, therefore, $\boxed{\hat{\mathbf{n}}_1 . \hat{\mathbf{n}}_2 = 0}$.

Worked Example 39

Find the cosine of the acute angle between the two planes whose equations are

(a) $\mathbf{r} \cdot (2\mathbf{i} - 3\mathbf{j} + 7\mathbf{k}) = 1$ and $\mathbf{r} \cdot (\mathbf{i} + \mathbf{j} + 2\mathbf{k}) = 2$

(b) $\mathbf{r} \cdot (\mathbf{i} - 3\mathbf{k}) = 5$ and $\mathbf{r} \cdot (2\mathbf{i} + \mathbf{j} - \mathbf{k}) = 3$.

Solution 39

(a) $\cos \theta = \hat{\mathbf{n}}_1 . \hat{\mathbf{n}}_2$

$= \dfrac{(2\mathbf{i} - 3\mathbf{j} + 7\mathbf{k})}{\sqrt{2^2 + 3^2 + 7^2}} \cdot \dfrac{(\mathbf{i} + \mathbf{j} + 2\mathbf{k})}{\sqrt{1 + 1 + 4}}$

$= \dfrac{2 - 3 + 14}{\sqrt{62}\sqrt{6}} = \dfrac{13}{\sqrt{372}} = 0.674$ to 3 d.p.

(b) $\cos \theta = \hat{\mathbf{n}}_1 . \hat{\mathbf{n}}_2$

$= \dfrac{\mathbf{i} - 3\mathbf{k}}{\sqrt{1 + 3^2}} \cdot \dfrac{2\mathbf{i} + \mathbf{j} - \mathbf{k}}{\sqrt{2^2 + 1^2 + 1^2}}$

$= \dfrac{2 + 3}{\sqrt{10}\sqrt{6}} = \dfrac{5}{\sqrt{60}} = 0.645$ to 3 d.p.

The Angle between a Line and a Plane

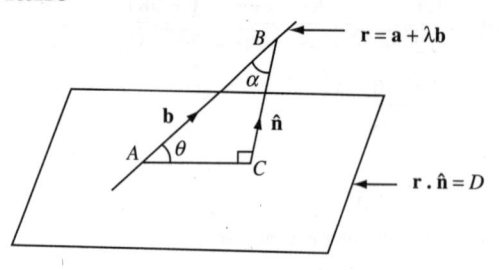

Fig. 8-I/45 The angle between a line and a plane

Consider the vector equation of a line to be $\mathbf{r} = \mathbf{a} + \lambda\mathbf{b}$ and the plane $\mathbf{r} \cdot \hat{\mathbf{n}} = D$.

Let α be the angle between the line and the normal to the plane

$$\cos \alpha = \frac{\mathbf{b} \cdot \hat{\mathbf{n}}}{|\mathbf{b}|}$$

If θ is the angle between the line and the plane,

$$\theta + \alpha = \frac{\Pi}{2}$$

$$\cos \alpha = \frac{BC}{AB} = \sin \theta$$

$$\boxed{\sin \theta = \frac{\mathbf{b} \cdot \hat{\mathbf{n}}}{|\mathbf{b}|}}$$

WORKED EXAMPLE 40

Determine the angle between the line with vector equation $\mathbf{r} = (2\mathbf{i} + \mathbf{j} - \mathbf{k}) + \lambda(2\mathbf{i} + 3\mathbf{j} + \mathbf{k})$ and the plane $\mathbf{r} \cdot (3\mathbf{i} + 4\mathbf{j} + 5\mathbf{k}) = 7$.

Solution 40

$$\sin \theta = \frac{\mathbf{b} \cdot \hat{\mathbf{n}}}{|\mathbf{b}|}$$

The direction vector $\mathbf{b} = 2\mathbf{i} + 3\mathbf{j} + \mathbf{k}$

$$|\mathbf{b}| = \sqrt{4 + 9 + 1} = \sqrt{14}$$

$$\hat{\mathbf{n}} = \frac{3\mathbf{i} + 4\mathbf{j} + 5\mathbf{k}}{\sqrt{3^2 + 4^2 + 5^2}}$$

$$= \frac{3}{\sqrt{50}}\mathbf{i} + \frac{4}{\sqrt{50}}\mathbf{j} + \frac{5}{\sqrt{50}}\mathbf{k}$$

$$\sin \theta = \frac{\mathbf{b} \cdot \hat{\mathbf{n}}}{\sqrt{14}}$$

$$= \frac{(2\mathbf{i} + 3\mathbf{j} + \mathbf{k}) \cdot \left(\frac{3}{\sqrt{50}}\mathbf{i} + \frac{4}{\sqrt{50}}\mathbf{j} + \frac{5}{\sqrt{50}}\mathbf{k}\right)}{\sqrt{14}}$$

$$= \frac{1}{\sqrt{50}\sqrt{14}}(6 + 12 + 5) = \frac{23}{\sqrt{700}} = 0.869318787$$

$$\theta = 60.4° \text{ to 3 s.f.}$$

WORKED EXAMPLE 41

Find the sine of the angle between the line and plane whose equations are

(a) $\mathbf{r} = (\mathbf{i} + 2\mathbf{j} - 5\mathbf{k}) + \lambda(2\mathbf{i} - 3\mathbf{j} + 4\mathbf{k})$,
    $\mathbf{r} \cdot (2\mathbf{i} - 3\mathbf{j} + 5\mathbf{k}) = 1$

(b) $\mathbf{r} = (-2\mathbf{i} + 3\mathbf{j} + 7\mathbf{k}) + \lambda(+4\mathbf{i} + \mathbf{j} + \mathbf{k})$,
    $\mathbf{r} \cdot (3\mathbf{i} + \mathbf{j} + \mathbf{k}) = 5$.

Solution 41

(a) $\sin \theta = \dfrac{\mathbf{b} \cdot \hat{\mathbf{n}}}{|\mathbf{b}|}$ $\quad \mathbf{b} = 2\mathbf{i} - 3\mathbf{j} + 4\mathbf{k}$

$$|\mathbf{b}| = \sqrt{2^2 + 3^2 + 4^2} = \sqrt{29}$$

$$\hat{\mathbf{n}} = \frac{2\mathbf{i} - 3\mathbf{j} + 5\mathbf{k}}{\sqrt{2^2 + 3^2 + 5^2}} = \frac{1}{\sqrt{38}}(2\mathbf{i} - 3\mathbf{j} + 5\mathbf{k})$$

$$\sin \theta = \frac{(2\mathbf{i} - 3\mathbf{j} + 4\mathbf{k}) \cdot (2\mathbf{i} - 3\mathbf{j} + 5\mathbf{k})}{\sqrt{29}\sqrt{38}}$$

$$= \frac{4 + 9 + 20}{\sqrt{29}\sqrt{38}}$$

$$\sin \theta = 0.994 \text{ to 3 d.p.}$$

(b) $\sin \theta = \dfrac{\mathbf{b} \cdot \hat{\mathbf{n}}}{|\mathbf{b}|} = \dfrac{(+4\mathbf{i} + \mathbf{j} + \mathbf{k})}{\sqrt{18}} \cdot \dfrac{(3\mathbf{i} + \mathbf{j} + \mathbf{k})}{\sqrt{11}}$

$$= \frac{12 + 1 + 1}{\sqrt{198}} = \frac{14}{\sqrt{198}}$$

$$= 0.995 \text{ to 3 d.p.}$$

The Intersection of Two Planes is a Line

Let two planes Π_1 and Π_2 with equations $\mathbf{r} \cdot \hat{\mathbf{n}}_1 = D_1$ and $\mathbf{r} \cdot \hat{\mathbf{n}}_2 = D_2$ respectively, intersect in a line.

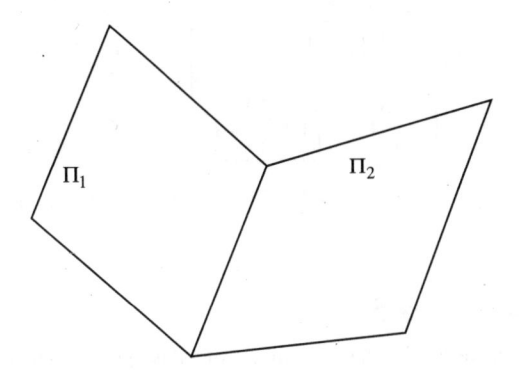

Fig. 8-I/46 The intersection of two planes.

The equation of the plane passing through the intersection has an equation

$\mathbf{r} \cdot (\hat{\mathbf{n}}_1 - k\hat{\mathbf{n}}_2) = D_1 - kD_2$.

WORKED EXAMPLE 42

Find the equation of the line of intersection of the two planes with equations

$\Pi_1 : \mathbf{r} \cdot (2\mathbf{i} - \mathbf{j} + 2\mathbf{k}) = 3$

$\Pi_2 : \mathbf{r} \cdot (-3\mathbf{i} + 2\mathbf{j} - \mathbf{k}) = 5$.

Solution 42

Let $\mathbf{r} = x\mathbf{i} + y\mathbf{j} + z\mathbf{k}$, $(x\mathbf{i} + y\mathbf{j} + z\mathbf{k}) \cdot (2\mathbf{i} - \mathbf{j} + 2\mathbf{k}) = 3$

$\boxed{2x - y + 2z = 3}$ ...(1)

$(x\mathbf{i} + y\mathbf{j} + z\mathbf{k}) \cdot (-3\mathbf{i} + 2\mathbf{j} - \mathbf{k}) = 5$

$\boxed{-3x + 2y - z = 5}$ ...(2)

Eliminating y from (1) and (2)

$4x - 2y + 4z = 6$
$\underline{-3x + 2y - z = 5}$
$x + 3z = 11$
$x = 11 - 3z$.

Eliminating z from (1) and (2)

$2x - y + 2z = 3$
$\underline{-6x + 4y - 2z = 10}$
$-4x + 3y = 13$

$x = \frac{3}{4}y - \frac{13}{4}$

$x = 11 - 3z = \frac{3}{4}y - \frac{13}{4} = \lambda$

$x = \lambda, \ \frac{11 - \lambda}{3} = z, \ y = \frac{4}{3}\lambda + \frac{13}{3}$.

Any point on the line of the intersection is

$\left(\lambda, \frac{4}{3}\lambda + \frac{13}{3}, \frac{11-\lambda}{3}\right)$ or $\left(3\mu, 4\mu + \frac{13}{3}, \frac{11}{3} - \mu\right)$

where $\lambda = 3\mu$.

The position vector therefore, of any point on the line is given

$\boxed{\mathbf{r} = \frac{13}{3}\mathbf{j} + \frac{11}{3}\mathbf{k} + \mu(3\mathbf{i} + 4\mathbf{j} - 1\mathbf{k})}$

The Parametric Form for the Vector Equation of a Plane

The plane, through the point with position vector **a** and parallel to **b** and **c**, has equation

$\boxed{\mathbf{r} = \mathbf{a} + s\mathbf{b} + t\mathbf{c}}$

where s and t are independent parameters.

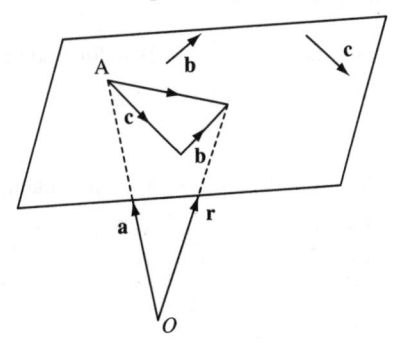

Fig. 8-I/47 The parametric form for the vector equation of a plane.

Fig. 8-I/47 shows a plane through the point A with position vector **a** and parallel to the direction vectors **b** and **c** as shown.

WORKED EXAMPLE 43

Two planes with parametric equations

$\Pi_1 : \mathbf{r} = (2\mathbf{i} - 3\mathbf{j} + \mathbf{k}) + s(-\mathbf{i} + 2\mathbf{j} + 3\mathbf{k})$
$\qquad + t(3\mathbf{i} + \mathbf{j} + 4\mathbf{k})$

$\Pi_2 : \mathbf{r} = (-\mathbf{i} + 2\mathbf{j} - 4\mathbf{k}) + \lambda(\mathbf{i} + \mathbf{j} + \mathbf{k})$
$\qquad + \mu(-3\mathbf{i} + 3\mathbf{j} + 2\mathbf{k})$

intersect find the vector equation of the line of intersection.

Solution 43

$\mathbf{r} = (2\mathbf{i}-3\mathbf{j}+\mathbf{k})+s(-\mathbf{i}+2\mathbf{j}+3\mathbf{k})+t(3\mathbf{i}+\mathbf{j}+4\mathbf{k})$ ...(a)

$$\boxed{\begin{aligned}\mathbf{r} = &(2 - s + 3t)\mathbf{i} + (-3 + 2s + t)\mathbf{j} \\ &+ (1 + 3s + 4t)\mathbf{k}\end{aligned}} \quad \text{...(1)}$$

$\mathbf{r} = (-\mathbf{i}+2\mathbf{j}-4\mathbf{k})+\lambda(\mathbf{i}+\mathbf{j}+\mathbf{k})+\mu(-3\mathbf{i}+3\mathbf{j}+2\mathbf{k})$ ...(b)

$$\boxed{\begin{aligned}\mathbf{r} = &(-1 + \lambda - 3\mu)\mathbf{i} + (2 + \lambda + 3\mu)\mathbf{j} \\ &+ (-4 + \lambda + 2\mu)\mathbf{k}\end{aligned}} \quad \text{...(2)}$$

These planes meet when

$2 - s + 3t = -1 + \lambda - 3\mu$ ...(3)

$-3 + 2s + t = 2 + \lambda + 3\mu$ ...(4)

$1 + 3s + 4t = -4 + \lambda + 2\mu$ ...(5)

Eliminating the two parameters s and t.

Eliminate firstly s

(3) × 2 $4 - 2s + 6t = -2 + 2\lambda - 6\mu$ adding

(4) $\underline{-3 + 2s + t = 2 + \lambda + 3\mu}$

$1 + 7t = 3\lambda - 3\mu$ ...(6)

(3) × 3 $6 - 3s + 9t = -3 + 3\lambda - 9\mu$ adding

(5) $\underline{1 + 3s + 4t = -4 + \lambda + 2\mu}$

$7 + 13t = -7 + 4\lambda - 7\mu$ ...(7)

Eliminate secondly t

(6) × 13 $13 + 91t = 39\lambda - 39\mu$ adding

(7) × (−7) $\underline{-49 - 91t = 49 - 28\lambda + 49\mu}$

$-36 = 49 + 11\lambda + 10\mu$

$11\lambda + 10\mu = -85 \quad \lambda = -\dfrac{10}{11}\mu - \dfrac{85}{11}$

substituting in (b)

$\mathbf{r} = (-\mathbf{i} + 2\mathbf{j} - 4\mathbf{k}) + \left(-\dfrac{10}{11}\mu - \dfrac{85}{11}\right)(\mathbf{i}+\mathbf{j}+\mathbf{k})$

$\quad + \mu(-3\mathbf{i} + 3\mathbf{j} + 2\mathbf{k})$

$\mathbf{r} = (-\mathbf{i} + 2\mathbf{j} - 4\mathbf{k}) - \dfrac{10}{11}\mu(\mathbf{i}+\mathbf{j}+\mathbf{k}) - \dfrac{85}{11}(\mathbf{i}+\mathbf{j}+\mathbf{k})$

$\quad + \mu(-3\mathbf{i} + 3\mathbf{j} + 2\mathbf{k})$

$$\boxed{\begin{aligned}\mathbf{r} = &\left(-\dfrac{96}{11}\mathbf{i} - \dfrac{63}{11}\mathbf{j} - \dfrac{129}{11}\mathbf{k}\right) \\ &+ \mu\left(-\dfrac{43}{11}\mathbf{i} + \dfrac{23}{11}\mathbf{j} + \dfrac{12}{11}\mathbf{k}\right)\end{aligned}}$$

the vector equation of the line of intersection of the planes.

Exercises 5

1. Determine the equation of the plane when $A(-1, -2, -3)$ is a point on the plane which is perpendicular to the line with direction ratios $2 : 3 : 4$.

2. Determine the equation of the plane when $A(2, -3, 4)$ is a point on the plane which is perpendicular to the line with direction ratios $l : m : n$.

3. Find the equation of the plane which contains three fixed points: $A(-3, 4, 7)$, $B(0, -2, 5)$ and $C(2, 0, -3)$.

4. Find the equation of the plane which contains three fixed points: $A(0, 0, 0), B(1, -2, -3), C(-2, 1, 2)$.

5. Find the equations of the planes passing through the fixed points and whose normals have the given direction ratios

   (i) $A(0, 0, 1)$, $1 : -2 : 3$

   (ii) $B(1, 0, 0)$, $4 : -5 : 6$

   (iii) $C(1, 3, 0)$, $-1 : 2 : 4$.

6. Find the equations of the planes passing through the set of points.

   (i) $A(1, 1, 0)$, $B(2, -2, 3)$, $C(0, 0, 2)$

   (ii) $P(0, 1, 0)$, $Q(-1, 3, -4)$, $R(1, 0, 2)$

   (iii) $D(-1, -2, -3)$, $E(0, 3, 0)$, $F(1, 2, 4)$.

7. Find the point of intersection of the line $l : \mathbf{r} = (2\mathbf{i} - 3\mathbf{j} + \mathbf{k}) + \lambda(-3\mathbf{i} + \mathbf{j} - 3\mathbf{k})$ and the plane $\Pi : \mathbf{r} \cdot (\mathbf{i} + \mathbf{j} + \mathbf{k}) = 3$.

8. Find the point of intersection of the line $l : \mathbf{r} = (-\mathbf{i} + 4\mathbf{j} - 5\mathbf{k}) + t(2\mathbf{i} - \mathbf{j} - \mathbf{k})$ and the plane $\Pi : \mathbf{r} \cdot (2\mathbf{i} - 2\mathbf{j} + 4\mathbf{k}) = 5$.

9. Find the point of intersection of the line $l : \mathbf{r} = (\mathbf{i} + \mathbf{j} + \mathbf{k}) + \mu(2\mathbf{i} - 3\mathbf{j} + 4\mathbf{k}) = 1$ and the plane $\Pi : \mathbf{r} \cdot (2\mathbf{i} + 5\mathbf{j} - 7\mathbf{k}) = 2$.

10. Find the point of intersection of the line $\dfrac{x+1}{2} = \dfrac{y-3}{3} = \dfrac{z+2}{4} = \lambda$ and the plane $x + y + 2z = 4$.

11. Find the point of intersection of the line $x = y = z = \lambda$ and the plane $\mathbf{r} \cdot (\mathbf{i} + 2\mathbf{j} + 3\mathbf{k}) = 7$.

12. The vector equations of a line and a plane are given
    $l : \mathbf{r} = 2\mathbf{i} - 3\mathbf{j} + \mathbf{k} + \lambda(-2\mathbf{i} + 5\mathbf{j} + \mathbf{k})$
    $\Pi : \mathbf{r} \cdot (-2\mathbf{i} - \mathbf{j} + \mathbf{k}) = 2$ respectively. Show that the line, l and the plane Π are parallel.

13. The vector equations of a line and a plane are given
    $l : \mathbf{r} = (-\mathbf{i} + 2\mathbf{j} - 5\mathbf{k}) + \mu(\mathbf{i} - 7\mathbf{j} + 4\mathbf{k})$
    $\Pi : \mathbf{r} \cdot (3\mathbf{i} + \mathbf{j} + \mathbf{k}) = 5$ respectively. Show that the line, l and the plane, Π are parallel.

14. Show that the line l whose vector equation is $\mathbf{r} = (\mathbf{i} - 3\mathbf{j} + 4\mathbf{k}) + t(-5\mathbf{i} + 7\mathbf{j} - 8\mathbf{k})$ is parallel to the plane Π whose vector equation is $\mathbf{r} \cdot (-2\mathbf{i} + 2\mathbf{j} + 3\mathbf{k}) = 1$ and find the distance between the line l and the plane Π.

15. Determine whether the following lines
    (i) $\mathbf{r} = \mathbf{i} + \mathbf{j} + \mathbf{k} + \lambda(-3\mathbf{i} + 4\mathbf{j} - 5\mathbf{k})$
    (ii) $\mathbf{r} = \mathbf{j} + 2\mathbf{k} + t(2\mathbf{i} - 3\mathbf{j} + \mathbf{k})$
    (iii) $\mathbf{r} = 2\mathbf{i} - 3\mathbf{j} - 7\mathbf{k} + \mu(3\mathbf{i} + 5\mathbf{j} + 9\mathbf{k})$.
    are parallel to the plane $\mathbf{r} \cdot (2\mathbf{i} - 3\mathbf{j} + \mathbf{k}) = d$ and find the acute angle between the line and the plane.

16. Find the vector equations of the following planes in the form $\mathbf{r} \cdot \mathbf{n} = d$:
    (i) $\mathbf{r} = \mathbf{i} + \mathbf{j} + \mathbf{k} + \lambda(2\mathbf{i} - \mathbf{j} + 3\mathbf{k}) + s(-2\mathbf{i} + 3\mathbf{j} - 7\mathbf{k})$
    (ii) $\mathbf{r} = 2\mathbf{j} - 3\mathbf{k} + \mu(-3\mathbf{i} + 4\mathbf{j} - \mathbf{k}) + \nu(2\mathbf{i} + 5\mathbf{j} - 5\mathbf{k})$
    (iii) $\mathbf{r} = (1 - t - s)\mathbf{a} + t\mathbf{b} + s\mathbf{c}$
    where $\mathbf{a} = 2\mathbf{i} - 3\mathbf{j} - \mathbf{k}$, $\mathbf{b} = -3\mathbf{i} + 4\mathbf{j} + \mathbf{k}$, $\mathbf{c} = 3\mathbf{j} - 5\mathbf{k}$ are position vectors.

17. Find the cartesian equations of the following planes:
    (i) $\mathbf{r} \cdot (2\mathbf{i} + 3\mathbf{j} + 4\mathbf{k}) = 3$
    (ii) $\mathbf{r} \cdot (-3\mathbf{i} + 2\mathbf{j} - \mathbf{k}) = 1$
    (iii) $\mathbf{r} \cdot (\mathbf{i} - 2\mathbf{j} + 3\mathbf{k}) = 0$.

18. Find the vector equation in the form $\mathbf{r} \cdot \mathbf{n} = d$ of the planes containing the following pairs of lines

    (i) $\mathbf{r} = (2\mathbf{i} - 3\mathbf{j} + 5\mathbf{k}) + \lambda(3\mathbf{j} - 4\mathbf{k})$,
       $\mathbf{r} = (2\mathbf{i} - 3\mathbf{j} + 5\mathbf{k}) + \mu(\mathbf{i} + 3\mathbf{k})$
    (ii) $\mathbf{r} = (2\mathbf{i} - 3\mathbf{k}) + \lambda(2\mathbf{i} - 3\mathbf{j})$,
       $\mathbf{r} = (2\mathbf{i} - 3\mathbf{k}) + \mu(3\mathbf{i} - \mathbf{j} - \mathbf{k})$
    (iii) $\mathbf{r} = (2\mathbf{i} + 7\mathbf{k}) + \lambda(\mathbf{i} + \mathbf{j} - \mathbf{k})$,
       $\mathbf{r} = (2\mathbf{i} + 7\mathbf{k}) + \mu(2\mathbf{i} - 3\mathbf{j} + \mathbf{k})$.

19. Find the vector equation of the plane containing the position vectors of three points $\mathbf{a} = 2\mathbf{j} - 3\mathbf{k}$, $\mathbf{b} = -2\mathbf{i} + 7\mathbf{j}$, $\mathbf{c} = (\mathbf{i} + \mathbf{j} + \mathbf{k})$.

20. Find the vector equation of the plane containing the three points $A(-1, -2, -3)$, $B(3, 4, 5)$, $C(-6, -7, 8)$.

21. Determine the distance between the two parallel planes
    $\Pi_1 : \mathbf{r} \cdot (\mathbf{i} + \mathbf{j} + \mathbf{k}) = 1$
    $\Pi_2 : \mathbf{r} \cdot (\mathbf{i} + \mathbf{j} + \mathbf{k}) = 2$.

22. Find the perpendicular distance of the point $(1, 2, 3)$ from the following planes:
    (i) $\mathbf{r} \cdot (3\mathbf{i} + 4\mathbf{j} + 5\mathbf{k}) = 12$
    (ii) $\mathbf{r} \cdot (6\mathbf{i} + 7\mathbf{j} + 8\mathbf{k}) = 5$
    (iii) $\mathbf{r} \cdot (\mathbf{i} + \mathbf{j} + \mathbf{k}) = 1$.

23. Two right angled cones have their bases contained in the planes $\mathbf{r} \cdot (\mathbf{i} + 2\mathbf{j} + 3\mathbf{k}) = 4$, $\mathbf{r} \cdot (\mathbf{i} + 2\mathbf{j} + 3\mathbf{k}) = 7$, determine the distance between the circular bases.

24. Show that the point $(-3, 4, -5)$ lies in the plane Π whose equation is $\mathbf{r} \cdot (2\mathbf{i} - 3\mathbf{j} - 12\mathbf{k}) = 42$.

25. Find the sine of the angle between the line and the plane whose equation are
    (i) $\mathbf{r} = 2\mathbf{j} + 3\mathbf{k} + \lambda(2\mathbf{i} + \mathbf{j} - 3\mathbf{k})$,
        $\mathbf{r} \cdot (\mathbf{i} + \mathbf{j} + \mathbf{k}) = 8$
    (ii) $\dfrac{x+1}{3} = \dfrac{y-1}{4} = \dfrac{z+2}{5} = \lambda$
        and $x + y + z = 3$.

26. A perpendicular line is drawn from the point $(1, 1, 1)$ to the plane $\mathbf{r} \cdot (-3\mathbf{i} + \mathbf{j} - 2\mathbf{k}) = 25$. Find the vector equation of the line and hence find the coordinates of the point of intersection.

27. The position vector of a point A is $-2\mathbf{i} + 3\mathbf{j} + 5\mathbf{k}$, a perpendicular line is drawn from A to intersect the

plane $3x - 5y - z = 5$ at B, find the coordinates of B and the reflection point C of A.

28. The position vectors of three points A, B and C are given:

    $\mathbf{a} = \mathbf{i} - \mathbf{j} - \mathbf{k}$, $\mathbf{b} = 2\mathbf{i} + 5\mathbf{j} + 7\mathbf{k}$,
    $\mathbf{c} = -\mathbf{i} - 2\mathbf{j} + 3\mathbf{k}$.

    Find the vectors $\overrightarrow{AB}$ and $\overrightarrow{AC}$ and hence find the vector equation of the plane ABC, repeat to procedure by finding the vectors $\overrightarrow{BC}$ and $\overrightarrow{BA}$ and hence find the vector equation of the plane ABC, in the form $\mathbf{r} \cdot \mathbf{n} = d$.

29. Show that the angle between a line with vector equation $\mathbf{r} = \mathbf{a} + \lambda \mathbf{b}$ and a plane with vector equation $\mathbf{r} \cdot \hat{\mathbf{n}} = d$ is given by

    $$\sin \theta = \frac{\mathbf{b} \cdot \hat{\mathbf{n}}}{|\mathbf{b}|}.$$

30. The position vector of two points A and B are given as $\mathbf{a} = 2\mathbf{i} - 3\mathbf{j} + 4\mathbf{k}$ and $\mathbf{b} = -3\mathbf{i} + 2\mathbf{j} - \mathbf{k}$, find the vector equation of the line AB.

    Determine the sine of the acute angle between the line AB and the plane with vector equation $\mathbf{r} \cdot (-4\mathbf{i} + 5\lambda - 7\mathbf{k}) = 27$.

31. The vector equation of a line is given $\mathbf{r} = (2\mathbf{i} - 5\mathbf{j} - 9\mathbf{k}) + \lambda(-7\mathbf{i} + 2\mathbf{j} + 3\mathbf{k})$ and the vector equation of a plane is given $\mathbf{r} \cdot (\mathbf{i} + 2\mathbf{j} + 3\mathbf{k}) = 33$. Find the angle between the line and the plane.

32. A line is parallel to the vector $3\mathbf{i} - 5\mathbf{j} + 7\mathbf{k}$ and passes through the point $(2, 2, 2)$. Find the cosine of the angle of the line and the plane with vector equation $\mathbf{r} \cdot (-\mathbf{j} + 5\mathbf{k}) = 5$.

33. Find the vector equations of the planes in parametric form containing the following pairs of lines:

    (a) $\mathbf{r} = 2\mathbf{i} - 3\mathbf{j} + 4\mathbf{k} + \lambda(\mathbf{i} + 5\mathbf{j} - 7\mathbf{k})$,

    $\mathbf{r} = \mathbf{i} + \mathbf{j} + 9\mathbf{k} + \mu(-3\mathbf{i} + 4\mathbf{j} + 8\mathbf{k})$

    (b) $\mathbf{r} = \mathbf{i} + 5\mathbf{j} + 9\mathbf{k} + \lambda(\mathbf{i} + 3\mathbf{j} - 2\mathbf{k})$,

    $\mathbf{r} = 2\mathbf{i} + 5\mathbf{j} - \mathbf{k} + \mu(3\mathbf{j} + 5\mathbf{k})$.

6
Vector Products

Introduction of Vector Product (or Cross Product)

If **a** and **b** are vectors, the vector product is given by $\boxed{\mathbf{a} \times \mathbf{b} = (ab \sin \theta)\hat{\mathbf{n}}}$ in a direction perpendicular to the plane containing **a** and **b** in the sense of a right handed screw turned from **a** to **b**.

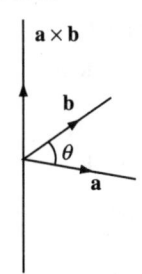

Fig. 8-I/48 Turned from **a** to **b**

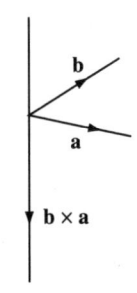

Fig. 8-I/49 Turned from **b** to **a**

The direction of $\mathbf{b} \times \mathbf{a}$ ($ab \sin \theta$) is the opposite to the direction $\mathbf{a} \times \mathbf{b}$.

$\boxed{\mathbf{a} \times \mathbf{b} = -\mathbf{b} \times \mathbf{a}}$.

The vector product is <u>not</u> commutative.

Let $\mathbf{a} = x_1\mathbf{i} + y_1\mathbf{j} + z_1\mathbf{k}$

$\mathbf{b} = x_2\mathbf{i} + y_2\mathbf{j} + z_2\mathbf{k}$.

To find the vector product

$\mathbf{a} \times \mathbf{b} = (x_1\mathbf{i} + y_1\mathbf{j} + z_1\mathbf{k}) \times (x_2\mathbf{i} + y_2\mathbf{j} + z_2\mathbf{k})$.

Consider the following parallel and perpendicular vectors: $\mathbf{i} \times \mathbf{i} = \mathbf{j} \times \mathbf{j} = \mathbf{k} \times \mathbf{k} = (1)(1)\sin 0 = 0$

$\boxed{\mathbf{i} \times \mathbf{i} = \mathbf{j} \times \mathbf{j} = \mathbf{k} \times \mathbf{k} = 0}$

therefore the vector product of parallel vectors is zero.

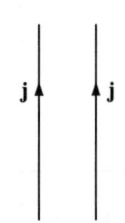

y-direction

Fig. 8-I/50

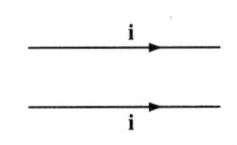

x-direction

Fig. 8-I/51

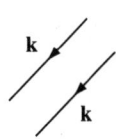

z-direction

Fig. 8-I/52

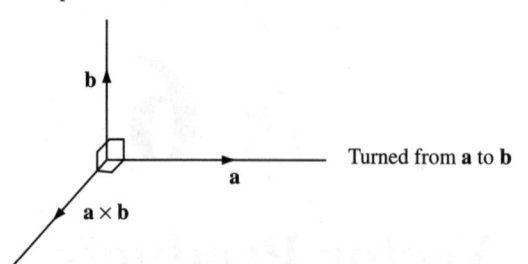

Perpendicular vector

Fig. 8-I/53
Turned from **a** to **b**

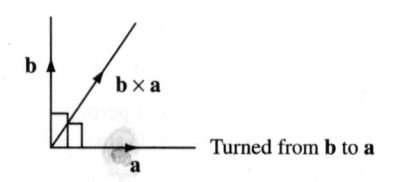

Fig. 8-I/54
Turned from **b** to **a**

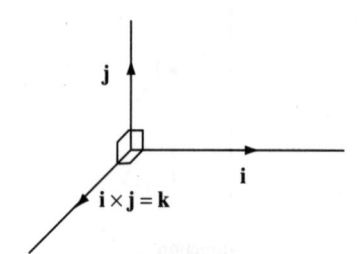

Fig. 8-I/55
$\mathbf{i} \times \mathbf{j} = \mathbf{k}$

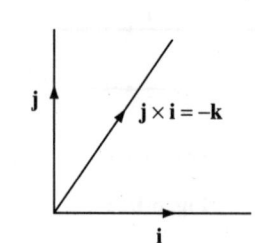

Fig. 8-I/56
$\mathbf{j} \times \mathbf{i} = -\mathbf{k}$

$\mathbf{i} \times \mathbf{j} = \mathbf{k} \quad \mathbf{j} \times \mathbf{i} = -\mathbf{k}$
$\mathbf{j} \times \mathbf{k} = \mathbf{i} \quad \mathbf{k} \times \mathbf{j} = -\mathbf{i}$
$\mathbf{k} \times \mathbf{i} = \mathbf{j} \quad \mathbf{i} \times \mathbf{k} = -\mathbf{j}$

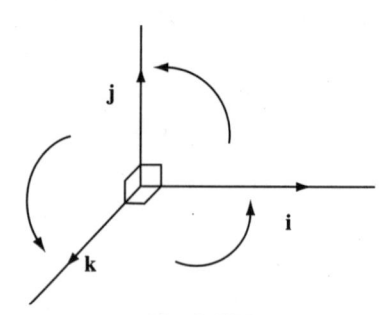

Fig. 8-I/57

Using the above and that the vector products are distributive we have

$$\mathbf{a} \times \mathbf{b} = (x_1\mathbf{i} + y_1\mathbf{j} + z_1\mathbf{k}) \times (x_2\mathbf{i} + y_2\mathbf{j} + z_2\mathbf{k})$$
$$= x_1x_2(\mathbf{i} \times \mathbf{i}) + x_1y_2(\mathbf{i} \times \mathbf{j}) + x_1z_2(\mathbf{i} \times \mathbf{k})$$
$$+ y_1x_2(\mathbf{j} \times \mathbf{i}) + y_1y_2(\mathbf{j} \times \mathbf{j}) + y_1z_2(\mathbf{j} \times \mathbf{k})$$
$$+ z_1x_2(\mathbf{k} \times \mathbf{i}) + z_1y_2(\mathbf{k} \times \mathbf{j}) + z_1z_2(\mathbf{k} \times \mathbf{k})$$
$$= x_1y_2\mathbf{k} - x_1z_2\mathbf{j} - y_1x_2\mathbf{k} + y_1z_2\mathbf{i}$$
$$+ z_1x_2\mathbf{j} - z_1y_2\mathbf{i}$$

$$\mathbf{a} \times \mathbf{b} = (y_1z_2 - z_1y_2)\mathbf{i} + (z_1x_2 - x_1z_2)\mathbf{j}$$
$$+ (x_1y_2 - y_1x_2)\mathbf{k} \text{ or}$$

$$\mathbf{a} \times \mathbf{b} = (y_1z_2 - z_1y_2)\mathbf{i} - (x_1z_2 - z_1x_2)\mathbf{j}$$
$$+ (x_1y_2 - y_1x_2)\mathbf{k}$$

this is rather tedious to obtain, it is observed that

$$\mathbf{a} \times \mathbf{b} = \begin{vmatrix} \mathbf{i} & \mathbf{j} & \mathbf{k} \\ x_1 & y_1 & z_1 \\ x_2 & y_2 & z_2 \end{vmatrix}$$
$$= \mathbf{i}(y_1z_2 - y_2z_1) - \mathbf{j}(x_1z_2 - x_2z_1)$$
$$+ \mathbf{k}(x_1y_2 - x_2y_1)$$

Vector Product

$$\mathbf{a} \times \mathbf{b} = (|a||b|\sin\theta)\hat{\mathbf{n}}$$
$$= (a_2b_3 - a_3b_2)\mathbf{i} + (a_3b_1 - a_1b_3)\mathbf{j}$$
$$+ (a_1b_2 - a_2b_1)\mathbf{k}$$

where $\hat{\mathbf{n}}$ is the unit vector perpendicular to both **a** and **b** such as **a**, **b**, **n** form a right-handed set.

WORKED EXAMPLE 44

Determine the vector products

(i) $\mathbf{a} \times \mathbf{b}$

(ii) $\mathbf{a} \times \mathbf{c}$

(iii) $\mathbf{b} \times \mathbf{c}$

where $\mathbf{a} = -\mathbf{i} + 2\mathbf{j} - 3\mathbf{k}$
$\mathbf{b} = 2\mathbf{i} + 3\mathbf{j} + 4\mathbf{k}$
$\mathbf{c} = 3\mathbf{i} - \mathbf{j} + 2\mathbf{k}$.

Solution 44

(i) $\mathbf{a} \times \mathbf{b} = \begin{vmatrix} \mathbf{i} & \mathbf{j} & \mathbf{k} \\ -1 & 2 & -3 \\ 2 & 3 & 4 \end{vmatrix}$

$= \mathbf{i}[2 \times 4 - (3)(-3)] - \mathbf{j}[(-1)(4) - (2)(-3)]$
$\quad + \mathbf{k}[(-1)(3) - 2 \times 2]$

$= 17\mathbf{i} - 2\mathbf{j} - 7\mathbf{k}$

(ii) $\mathbf{a} \times \mathbf{c} = \begin{vmatrix} \mathbf{i} & \mathbf{j} & \mathbf{k} \\ -1 & 2 & -3 \\ 3 & -1 & 2 \end{vmatrix}$

$= \mathbf{i}(4 - 3) - \mathbf{j}(-2 + 9) + \mathbf{k}(1 - 6)$

$= \mathbf{i} - 7\mathbf{j} - 5\mathbf{k}$

(iii) $\mathbf{b} \times \mathbf{c} = \begin{vmatrix} \mathbf{i} & \mathbf{j} & \mathbf{k} \\ 2 & 3 & 4 \\ 3 & -1 & 2 \end{vmatrix}$

$= \mathbf{i}(6 + 4) - \mathbf{j}(4 - 12) + \mathbf{k}(-2 - 9)$

$= 10\mathbf{i} + 8\mathbf{j} - 11\mathbf{k}$.

Observe that $\mathbf{a} \times \mathbf{b}, \mathbf{a} \times \mathbf{c}, \mathbf{b} \times \mathbf{c}$ are vectors although $\mathbf{a} \cdot \mathbf{b}, \mathbf{a} \cdot \mathbf{c}, \mathbf{b} \cdot \mathbf{c}$ are scalars. Therefore the cross products are vectors the dot products are scalars.

WORKED EXAMPLE 45

Simplify the following cross products.

(i) $\mathbf{c} \times (\mathbf{c} + \mathbf{b})$
(ii) $(\mathbf{a} \times \mathbf{a}) \times (\mathbf{a} + \mathbf{b})$
(iii) $(\mathbf{a} + \mathbf{b}) \times (\mathbf{a} - \mathbf{b})$
(iv) $(\mathbf{a} \times \mathbf{b}) \cdot \mathbf{a}$.

Solution 45

(i) $\mathbf{c} \times (\mathbf{c} + \mathbf{b}) = \mathbf{c} \times \mathbf{c} + \mathbf{c} \times \mathbf{b} = \mathbf{c} \times \mathbf{b}$
since $\mathbf{c} \times \mathbf{c} = 0$

(ii) $(\mathbf{a} \times \mathbf{a}) \times (\mathbf{a} + \mathbf{b}) = 0 \times (\mathbf{a} + \mathbf{b}) = 0$
since $\mathbf{a} \times \mathbf{a} = 0$

(iii) $(\mathbf{a} + \mathbf{b}) \times (\mathbf{a} - \mathbf{b}) = \mathbf{a} \times \mathbf{a} - \mathbf{a} \times \mathbf{b} + \mathbf{b} \times \mathbf{a} - \mathbf{b} \times \mathbf{b}$
$= -\mathbf{a} \times \mathbf{b} + \mathbf{b} \times \mathbf{a} = 2(\mathbf{b} \times \mathbf{a})$
since $\mathbf{a} \times \mathbf{a} = 0$ and $\mathbf{b} \times \mathbf{b} = 0$ but $\mathbf{a} \times \mathbf{b} = -\mathbf{b} \times \mathbf{a}$.

(iv) $\mathbf{a} \times \mathbf{b}$ is perpendicular to $\mathbf{a}$ and the scalar product $(\mathbf{a} \times \mathbf{b}) \cdot \mathbf{a}$ is zero since $|\mathbf{a} \times \mathbf{b}||\mathbf{a}| \cos 90° = 0$.

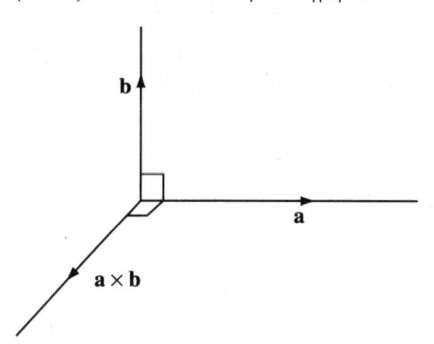

Fig. 8-I/58

WORKED EXAMPLE 46

If $\mathbf{a} = 2\mathbf{i} - 3\mathbf{j} + 5\mathbf{k}, \mathbf{b} = -3\mathbf{i} - 2\mathbf{k}, \mathbf{c} = \mathbf{i} + \mathbf{j} + \mathbf{k}$ find

(i) $(\mathbf{a} \times \mathbf{b}) \times \mathbf{c}$
(ii) $\mathbf{a} \times (\mathbf{b} \times \mathbf{c})$
(iii) $\mathbf{a} \cdot (\mathbf{b} \times \mathbf{c})$
(iv) $(\mathbf{a} \times \mathbf{b}) \cdot \mathbf{c}$
(v) $\mathbf{a} \cdot \mathbf{b} \cdot \mathbf{c}$.

Solution 46

(i) $\mathbf{a} \times \mathbf{b} = \begin{vmatrix} \mathbf{i} & \mathbf{j} & \mathbf{k} \\ 2 & -3 & 5 \\ -3 & 0 & -2 \end{vmatrix}$

$= \mathbf{i}(6) - \mathbf{j}(-4 + 15) + \mathbf{k}(-9)$

$= 6\mathbf{i} - 11\mathbf{j} - 9\mathbf{k}$

$(6\mathbf{i} - 11\mathbf{j} - 9\mathbf{k}) \times (\mathbf{i} + \mathbf{j} + \mathbf{k})$

$= \begin{vmatrix} \mathbf{i} & \mathbf{j} & \mathbf{k} \\ 6 & -11 & -9 \\ 1 & 1 & 1 \end{vmatrix}$

$= \mathbf{i}(-11 + 9) - \mathbf{j}(6 + 9) + \mathbf{k}(6 + 11)$

$= -2\mathbf{i} - 15\mathbf{j} + 17\mathbf{k}$

$(\mathbf{a} \times \mathbf{b}) \times \mathbf{c} = -2\mathbf{i} - 15\mathbf{j} + 17\mathbf{k}$

(ii) $\mathbf{a} \times (\mathbf{b} \times \mathbf{c})$

$\mathbf{b} \times \mathbf{c} = \begin{vmatrix} \mathbf{i} & \mathbf{j} & \mathbf{k} \\ -3 & 0 & -2 \\ 1 & 1 & 1 \end{vmatrix}$

$= \mathbf{i}(2) - \mathbf{j}(-3 + 2) + \mathbf{k}(-3)$

$= 2\mathbf{i} + \mathbf{j} - 3\mathbf{k}$

$$\mathbf{a} \times (\mathbf{b} \times \mathbf{c}) = \begin{vmatrix} \mathbf{i} & \mathbf{j} & \mathbf{k} \\ 2 & -3 & 5 \\ 2 & 1 & -3 \end{vmatrix}$$

$$= \mathbf{i}(9-5) - \mathbf{j}(-6-10) + \mathbf{k}(2+6)$$
$$= 4\mathbf{i} + 16\mathbf{j} + 8\mathbf{k}$$

(iii) $\mathbf{a} \cdot (\mathbf{b} \times \mathbf{c}) = \mathbf{a} \cdot (2\mathbf{i} + \mathbf{j} - 3\mathbf{k})$
$= (2\mathbf{i} - 3\mathbf{j} + 5\mathbf{k}) \cdot (2\mathbf{i} + \mathbf{j} - 3\mathbf{k})$
$= 4 - 3 - 15 = -14$

(iv) $(\mathbf{a} \times \mathbf{b}) \cdot \mathbf{c} = (6\mathbf{i} - 11\mathbf{j} - 9\mathbf{k}) \cdot (\mathbf{i} + \mathbf{j} + \mathbf{k})$
$= 6 - 11 - 9 = -14$

(v) $\mathbf{a} \cdot \mathbf{b} \cdot \mathbf{c} = [(2\mathbf{i} - 3\mathbf{j} + 5\mathbf{k}) \cdot (-3\mathbf{i} - 2\mathbf{k})](\mathbf{i} + \mathbf{j} + \mathbf{k})$
$= (-6 - 10) \cdot (\mathbf{i} + \mathbf{j} + \mathbf{k})$
$= -16(\mathbf{i} + \mathbf{j} + \mathbf{k}).$

WORKED EXAMPLE 47

Find the angle between the position vector $\overrightarrow{OP} = 3\mathbf{i} - 2\mathbf{j} + \mathbf{k}$ and $\overrightarrow{OQ} = -\mathbf{i} + \mathbf{j} + 4\mathbf{k}$ using the vector product formula.

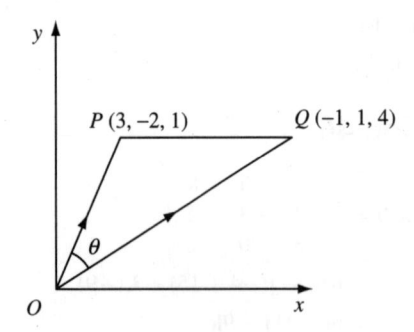

Fig. 8-I/59

Solution 47

$$\frac{|\overrightarrow{OP} \times \overrightarrow{OQ}|}{|\overrightarrow{OP}||\overrightarrow{OQ}|} = \sin \theta$$

$$\overrightarrow{OP} \times \overrightarrow{OQ} = (3\mathbf{i} - 2\mathbf{j} + \mathbf{k}) \times (-\mathbf{i} + \mathbf{j} + 4\mathbf{k})$$

$$= \begin{vmatrix} \mathbf{i} & \mathbf{j} & \mathbf{k} \\ 3 & -2 & 1 \\ -1 & 1 & 4 \end{vmatrix}$$

$$= \mathbf{i}(-8 - 1) - \mathbf{j}(12 + 1) + \mathbf{k}(3 - 2)$$
$$= -9\mathbf{i} - 13\mathbf{j} + \mathbf{k}$$

$$|\overrightarrow{OP} \times \overrightarrow{OQ}| = \sqrt{9^2 + 13^2 + 1^2} = 15.8 \text{ to 3 s.f.}$$

$$|\overrightarrow{OP}| = \sqrt{3^2 + 2^2 + 1^2} = 3.74 \text{ to 3 s.f.}$$

$$|\overrightarrow{OQ}| = \sqrt{1^2 + 1^2 + 4^2} = 4.24 \text{ to 3 s.f.}$$

$$\sin \theta = \frac{15.8}{3.74 \times 4.24} = 0.996367672$$

$\theta = 85.1°$.

Check that when rounding off only at the end, $\theta = 86.4°$.

The Perpendicular Distance, d, of a Point $P(x_1, y_1, z_1)$ from a Line with Vector Equation $r = \mathbf{a} + \lambda \mathbf{b}$

The position vectors of the points A and P are $\mathbf{a}$ and $\mathbf{p}$ respectively

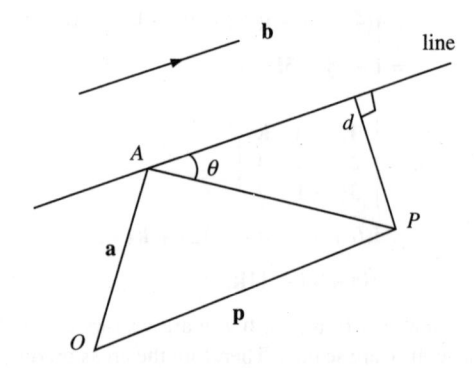

Fig. 8-I/60 The perpendicular distance from a line.

The vector product $= \mathbf{b} \times \overrightarrow{AP} = |\mathbf{b}| |\overrightarrow{AP}| \sin \theta \hat{\mathbf{n}}$

$$|\mathbf{b} \times \overrightarrow{AP}| = |\mathbf{b}| |\overrightarrow{AP}| \sin \theta$$

$$|\overrightarrow{AP}| \sin \theta = \frac{|\mathbf{b} \times \overrightarrow{AP}|}{|\mathbf{b}|} \quad \overrightarrow{AP} = \mathbf{p} - \mathbf{a}$$

$$\sin \theta = \frac{d}{\overrightarrow{AP}} \Rightarrow d = \overrightarrow{AP} \sin \theta$$

$$\boxed{d = \frac{|\mathbf{b} \times (\mathbf{p} - \mathbf{a})|}{|\mathbf{b}|}}$$

WORKED EXAMPLE 48

Find the perpendicular distance of the point $(-1, 2, -4)$ from the line whose vector equations is given
$\mathbf{r} = (2\mathbf{i} + \mathbf{j} - \mathbf{k}) + \lambda(-3\mathbf{i} + 4\mathbf{j} - 5\mathbf{k})$.

Solution 48

$\mathbf{b} = -3\mathbf{i} + 4\mathbf{j} - 5\mathbf{k}$ the direction vector

$\mathbf{p} = -\mathbf{i} + 2\mathbf{j} - 4\mathbf{k}$ the position vector of the point

$\mathbf{a} = 2\mathbf{i} + \mathbf{j} - \mathbf{k}$ the position vector of point A through the line

$\mathbf{p} - \mathbf{a} = -3\mathbf{i} + \mathbf{j} - 3\mathbf{k}$,
$\mathbf{b} = \sqrt{(-3)^2 + 4^2 + (-5)^2} = \sqrt{50} = 7.07$ to 3 s.f.

$\mathbf{b} \times (\mathbf{p} - \mathbf{a}) = (-3\mathbf{i} + 4\mathbf{j} - 5\mathbf{k}) \times (-3\mathbf{i} + \mathbf{j} - 3\mathbf{k})$

$$= \begin{vmatrix} \mathbf{i} & \mathbf{j} & \mathbf{k} \\ -3 & 4 & -5 \\ -3 & 1 & -3 \end{vmatrix}$$

$= \mathbf{i}(-12 + 5) - \mathbf{j}(9 - 15) + \mathbf{k}(-3 + 12)$

$= -7\mathbf{i} + 6\mathbf{j} + 9\mathbf{k}$,

$|\mathbf{b} \times (\mathbf{p} - \mathbf{a})| = \sqrt{49 + 36 + 81}$

$= \sqrt{166} = 12.9$ to 3 s.f.

$d = \dfrac{|\mathbf{b} \times (\mathbf{p} - \mathbf{a})|}{|\mathbf{b}|} = \dfrac{12.9}{7.07} = 1.82$ to 3 s.f.

WORKED EXAMPLE 49

A line passes through two points $A(1, 2, -3)$ and $B(-2, -3, 1)$ find the perpendicular distance from the origin to the line.

Solution 49

The vector equation of the line is $\mathbf{r} = (\mathbf{i} + 2\mathbf{j} - 3\mathbf{k}) + \lambda(-3\mathbf{i} - 5\mathbf{j} + 4\mathbf{k})$.

The position vector is $\mathbf{i} + 2\mathbf{j} - 3\mathbf{k}$ the direction vector is $-3\mathbf{i} - 5\mathbf{j} + 4\mathbf{k}$.

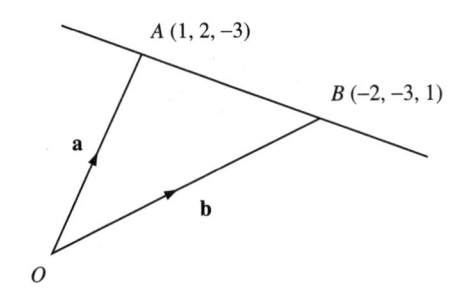

Fig. 8-I/61 Perpendicular distance from a point to the line

$d = \dfrac{|\mathbf{b} \times (\mathbf{p} - \mathbf{a})|}{(\mathbf{b})}$

where $d = \dfrac{\sqrt{75}}{\sqrt{50}} = 1.22$

$\mathbf{p} - \mathbf{a} = -\mathbf{i} - 2\mathbf{j} + 3\mathbf{k}$ where $\mathbf{p} = 0$ (origin),
$\mathbf{a} = \mathbf{i} + 2\mathbf{j} - 3\mathbf{k}$

$\mathbf{b} = -3\mathbf{i} - 5\mathbf{j} + 4\mathbf{k}$ the direction vector $= -2\mathbf{i} - 3\mathbf{j} + \mathbf{k} - (\mathbf{i} + 2\mathbf{j} - 3\mathbf{k}) = -3\mathbf{i} - 5\mathbf{j} + 4\mathbf{k}$

$|\mathbf{b}| = \sqrt{(-3)^2 + (-5)^2 + 4^2}$
$= \sqrt{9 + 25 + 16} = \sqrt{50}$

$\mathbf{b} \times (\mathbf{p} - \mathbf{a}) = (-3\mathbf{i} - 5\mathbf{j} + 4\mathbf{k}) \times (-\mathbf{i} - 2\mathbf{j} + 3\mathbf{k})$

$$= \begin{vmatrix} \mathbf{i} & \mathbf{j} & \mathbf{k} \\ -3 & -5 & 4 \\ -1 & -2 & 3 \end{vmatrix}$$

$= \mathbf{i}(-15 + 8) - \mathbf{j}(-9 + 4) + \mathbf{k}(6 - 5)$

$= -7\mathbf{i} + 5\mathbf{j} + \mathbf{k}$

$|\mathbf{b} \times (\mathbf{p} - \mathbf{a})| = \sqrt{(-7)^2 + 5^2 + 1^2} = \sqrt{75}$.

WORKED EXAMPLE 50

The coordinates of the vertices of a triangle ABC are given $A(-1, 2, 3), B(0, -3, 1), C(2, 0, -4)$. Determine (i) the angle between $\overrightarrow{AB}$ and $\overrightarrow{BC}$, (ii) the angle between AC and AB, (iii) the areas of the triangles (a) OAB (b) OBC (c) ABC.

Find also a unit vector which is perpendicular to the plane containing the triangle ABC.

Solution 50

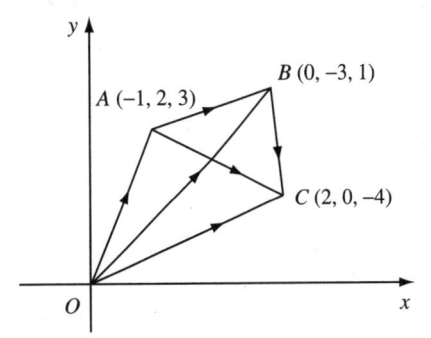

Fig. 8-I/62

(i) The position vectors of A, B, and C are:
$$\vec{OA} = -\mathbf{i} + 2\mathbf{j} + 3\mathbf{k}$$
$$\vec{OB} = -3\mathbf{j} + \mathbf{k}$$
$$\vec{OC} = 2\mathbf{i} - 4\mathbf{k}$$
$$\vec{AB} = \vec{OB} - \vec{OA} = -3\mathbf{j} + \mathbf{k} - (-\mathbf{i} + 2\mathbf{j} + 3\mathbf{k})$$
$$= \mathbf{i} - 5\mathbf{j} - 2\mathbf{k}$$
$$\vec{BC} = \vec{OC} - \vec{OB} = 2\mathbf{i} - 4\mathbf{k} - (-3\mathbf{j} + \mathbf{k})$$
$$= 2\mathbf{i} + 3\mathbf{j} - 5\mathbf{k}.$$

The vector product of $\vec{AB}$ and $\vec{BC}$ is given by
$$\left|\vec{AB} \times \vec{BC}\right| = \left|\vec{AB}\right|\left|\vec{BC}\right| \sin\theta$$

$$\sin\theta = \frac{\left|\vec{AB} \times \vec{BC}\right|}{\left|\vec{AB}\right|\left|\vec{BC}\right|} = \begin{vmatrix} \mathbf{i} & \mathbf{j} & \mathbf{k} \\ 1 & -5 & -2 \\ 2 & 3 & -5 \end{vmatrix}$$

$$= \mathbf{i}(25 + 6) - \mathbf{j}(-5 + 4) + \mathbf{k}(3 + 10)$$
$$= 31\mathbf{i} + \mathbf{j} + 13\mathbf{k}$$

$$\left|\vec{AB} \times \vec{BC}\right| = \sqrt{31^2 + 1^2 + 13^2} = 33.6 \text{ to 3 s.f.}$$

$$\left|\vec{AB}\right| = |\mathbf{i} - 5\mathbf{j} - 2\mathbf{k}|$$
$$= \sqrt{1^2 + 5^2 + 2^2} = 5.48 \text{ to 3 s.f.}$$

$$\left|\vec{BC}\right| = |2\mathbf{i} + 3\mathbf{j} - 5\mathbf{k}|$$
$$= \sqrt{2^2 + 3^2 + 5^2} = 6.16 \text{ to 3 s.f.}$$

$$\sin\theta = \frac{\left|\vec{AB} \times \vec{BC}\right|}{\left|\vec{AB}\right|\left|\vec{BC}\right|} = \frac{33.6}{5.48 \times 6.16}$$

$$= 0.99535501$$

$$\theta = 84.5° \text{ to 3 s.f.}$$

If we round off only at the end, $\theta = 84.9°$ to 3 s.f.

(ii) $\vec{AC} = \vec{OC} - \vec{OA} = 2\mathbf{i} - 4\mathbf{k} - (-\mathbf{i} + 2\mathbf{j} + 3\mathbf{k})$
$$= 3\mathbf{i} - 2\mathbf{j} - 7\mathbf{k}$$
$\vec{AB} = \mathbf{i} - 5\mathbf{j} - 2\mathbf{k}.$

The vector product of $\vec{AC}$ and $\vec{AB}$ is
$$\vec{AC} \times \vec{AB} = (3\mathbf{i} - 2\mathbf{j} - 7\mathbf{k}) \times (\mathbf{i} - 5\mathbf{j} - 2\mathbf{k})$$

$$= \begin{vmatrix} \mathbf{i} & \mathbf{j} & \mathbf{k} \\ 3 & -2 & -7 \\ 1 & -5 & -2 \end{vmatrix}$$

$$= \mathbf{i}(4 - 35) - \mathbf{j}(-6 + 7) + \mathbf{k}(-15 + 2)$$
$$= -31\mathbf{i} - \mathbf{j} - 13\mathbf{k}$$

$$\left|\vec{AC} \times \vec{AB}\right| = \sqrt{31^2 + 1^2 + 13^2} = 33.6$$

$\vec{AC} = \vec{OC} - \vec{OA} = 2\mathbf{i} - 4\mathbf{k} - (-\mathbf{i} + 2\mathbf{j} + 3\mathbf{k})$
$$= 3\mathbf{i} - 2\mathbf{j} - 7\mathbf{k}$$

$$\left|\vec{AC}\right| = \sqrt{3^2 + 2^2 + 7^2} = 7.87$$

$$\left|\vec{AB}\right| = \sqrt{1 + 5^2 + 2^2} = 5.48$$

$$\sin\phi = \frac{\left|\vec{AC} \times \vec{AB}\right|}{\left|\vec{AC}\right|\left|\vec{AB}\right|} = \frac{33.6}{(7.87)(5.48)} = 0.7791$$

$$\phi = 51.2° \text{ to 3 s.f.}$$

If we round off only at the end, $\phi = 51.2°$ to 3 s.f.

(iii) (a) Area $\triangle OAB = \sqrt{s(s-a)(s-b)(s-c)}$

$$\left|\vec{OA}\right| = \sqrt{1^2 + 2^2 + 3^2} = 3.74$$

$$\left|\vec{OB}\right| = \sqrt{(-3)^2 + 1^2} = 3.16$$

$$\left|\vec{AB}\right| = 5.48$$

$$s = \frac{3.74 + 3.16 + 5.48}{2} = 6.19$$

Area $\triangle OAB$
$$= \sqrt{6.19(6.19 - 3.74)(6.19 - 3.16)(6.19 - 5.48)}$$
$$= \sqrt{6.19 \times 2.45 \times 3.03 \times 0.71}$$
$$= 5.71 \text{ s.u. to 3 s.f.}$$

(b) Area $\triangle OBC = \sqrt{s(s-a)(s-b)(s-c)}$

$$\left|\vec{OB}\right| = 3.16,$$

$$\left|\vec{OC}\right| = \sqrt{2^2 + 4^2} = \sqrt{20} = 4.47,$$

$$\left|\vec{BC}\right| = 6.16$$

$$s = \frac{3.16 + 4.47 + 6.16}{2} = 6.9$$

Area $\triangle OBC$
$$= \sqrt{\begin{array}{c} 6.9 \times (6.9 - 3.16) \times (6.9 - 4.47) \\ \times (6.9 - 6.15) \end{array}}$$
$$= \sqrt{6.9 \times 3.74 \times 2.43 \times 0.75}$$
$$= 6.86 \text{ s.u. to 3 s.f.}$$

(c) Area $\triangle ABC$

$$= \sqrt{\begin{array}{c}6.9 \times (6.9 - 3.16) \times (6.9 - 4.47)\\ \times (6.9 - 6.15)\end{array}}$$

$s = |\overrightarrow{AB}| = 5.48$, $|\overrightarrow{BC}| = 6.16$,

$|\overrightarrow{AC}| = 7.87$

$s = \dfrac{5.48 + 6.16 + 7.87}{2} = 9.76$

Area $\triangle ABC$

$$= \sqrt{\begin{array}{c}9.76 \times (9.76 - 5.48) \times (9.76 - 6.16)\\ \times (9.76 - 7.87)\end{array}}$$

$= \sqrt{9.76 \times 4.28 \times 3.6 \times 1.89}$

$= 16.9$ s.u. to 3 s.f.

To find the unit vector which is perpendicular to the plane containing the triangle ABC.

$\overrightarrow{AB} \times \overrightarrow{BC}$ is a vector which is perpendicular to both $\overrightarrow{AB}$ and $\overrightarrow{BC}$ and is therefore perpendicular to the plane of the triangle ABC.

$\hat{\mathbf{n}} = \dfrac{\overrightarrow{AB} \times \overrightarrow{BC}}{|\overrightarrow{AB} \times \overrightarrow{BC}|} = \dfrac{31\mathbf{i} + \mathbf{j} + 13\mathbf{k}}{\sqrt{31^2 + 1^2 + 13^2}}$

$= \dfrac{1}{33.6}(31\mathbf{i} + \mathbf{j} + 13\mathbf{k})$

$\hat{\mathbf{n}} = \dfrac{\overrightarrow{AB} \times \overrightarrow{BC}}{|\overrightarrow{AB} \times \overrightarrow{BC}|} = \dfrac{-31\mathbf{i} - \mathbf{j} - 13\mathbf{k}}{33.6}$

$= -\dfrac{1}{33.6}(31\mathbf{i} + \mathbf{j} + 13\mathbf{k})$

this is equal and opposite to the unit vector we found previously.

Area of Triangle

Fig. 8-I/63 $\sin\theta = \dfrac{h}{|\overrightarrow{AB}|}$

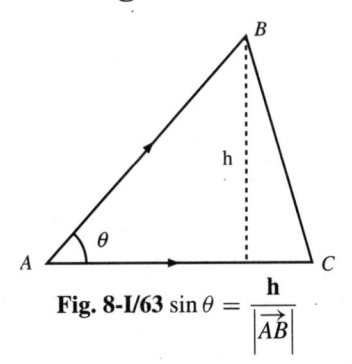

The area of $\triangle ABC = \dfrac{1}{2}$ base $\times$ height

$= \dfrac{1}{2}|\overrightarrow{AC}| \times h$

$= \dfrac{1}{2}|\overrightarrow{AC}| \times \left(|\overrightarrow{AB}|\sin\theta\right)$

$= \dfrac{1}{2}|\overrightarrow{AC} \times \overrightarrow{AB}|$.

From the previous example

Area $\triangle OAB = \dfrac{1}{2}|\overrightarrow{OA} \times \overrightarrow{OB}|$

$= \dfrac{1}{2}(11.45) = 5.71$ s.u. to 3 s.f.

since

$\overrightarrow{OA} \times \overrightarrow{OB} = \begin{vmatrix} \mathbf{i} & \mathbf{j} & \mathbf{k} \\ -1 & 2 & 3 \\ 0 & -3 & 1 \end{vmatrix}$

$= \mathbf{i}(2+9) - \mathbf{j}(-1) + \mathbf{k}(3)$

$= 11\mathbf{i} + \mathbf{j} + 3\mathbf{k}$

$|\overrightarrow{OA} \times \overrightarrow{OB}| = \sqrt{11^2 + 1^2 + 3^2} = 11.45$

Area $\triangle OBC = \dfrac{1}{2}|\overrightarrow{OB} \times \overrightarrow{OC}|$

$= \dfrac{1}{2}(13.56)$

$= 6.78$ square units

since

$|\overrightarrow{OB}| \times |\overrightarrow{OC}| = \begin{vmatrix} \mathbf{i} & \mathbf{j} & \mathbf{k} \\ 0 & -3 & 1 \\ 2 & 0 & -4 \end{vmatrix}$

$= \mathbf{i}(12) - \mathbf{j}(-2) + \mathbf{k}(6)$

$= 12\mathbf{i} + 2\mathbf{j} + 6\mathbf{k}$

$|\overrightarrow{OB}| \times |\overrightarrow{OC}| = \sqrt{12^2 + 2^2 + 6^2} = 13.56$

Area $\triangle ABC = \dfrac{1}{2}|\overrightarrow{AB} \times \overrightarrow{BC}|$

$= \dfrac{1}{2}|31\mathbf{i} + \mathbf{j} + 13\mathbf{k}|$

$= \dfrac{1}{2} \times 33.6 = 16.8$ square units.

WORKED EXAMPLE 51

Two sides of a triangle are given by the vectors $\mathbf{a} = 2\mathbf{i} + 3\mathbf{j} - \mathbf{k}$, $\mathbf{b} = -\mathbf{i} + 4\mathbf{j} + 2\mathbf{k}$. Find the area of the triangle using the vector product.

Solution 51

Area $\Delta = \frac{1}{2}|\mathbf{a} \times \mathbf{b}|$

$= \frac{1}{2}|10\mathbf{i} - 3\mathbf{j} + 11\mathbf{k}|$

$= \frac{1}{2}\sqrt{10^2 + 3^2 + 11^2}$

$= \frac{15.16575089}{2} = 7.58$ s.u. to 3 s.f.

where $\mathbf{a} \times \mathbf{b} = \begin{vmatrix} \mathbf{i} & \mathbf{j} & \mathbf{k} \\ 2 & 3 & -1 \\ -1 & 4 & +2 \end{vmatrix}$

$= \mathbf{i}(6+4) - \mathbf{j}(4-1) + \mathbf{k}(8+3)$

$= 10\mathbf{i} - 3\mathbf{j} + 11\mathbf{k}.$

WORKED EXAMPLE 52

Prove the sine rule of a triangle ABC, $\dfrac{a}{\sin A} = \dfrac{b}{\sin B} = \dfrac{c}{\sin C}$ using vector products.

Solution 52

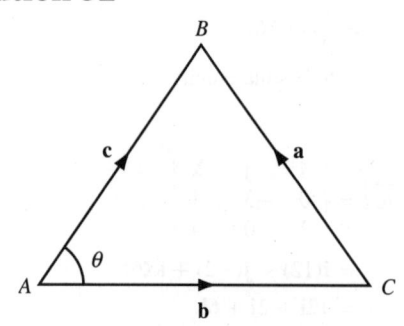

Fig. 8-I/64

Area of $\triangle ABC = \frac{1}{2}|\mathbf{a} \times \mathbf{b}| = \frac{1}{2}|\mathbf{a}||\mathbf{b}|\sin C$

Area of $\triangle ABC = \frac{1}{2}|\mathbf{b} \times \mathbf{c}| = \frac{1}{2}|\mathbf{b}||\mathbf{c}|\sin A$

Area of $\triangle ABC = \frac{1}{2}|\mathbf{a} \times \mathbf{c}| = \frac{1}{2}|\mathbf{a}||\mathbf{c}|\sin B$

$\frac{1}{2}|\mathbf{a}||\mathbf{b}|\sin C = \frac{1}{2}|\mathbf{b}||\mathbf{c}|\sin A = \frac{1}{2}|\mathbf{a}||\mathbf{c}|\sin B$

$|\mathbf{a}|\sin C = |\mathbf{c}|\sin A \qquad |\mathbf{b}|\sin A = |\mathbf{a}|\sin B$

$\dfrac{|\mathbf{a}|}{\sin A} = \dfrac{|\mathbf{c}|}{\sin C} \qquad \dfrac{|\mathbf{a}|}{\sin A} = \dfrac{|\mathbf{b}|}{\sin B}$

therefore $\dfrac{|\mathbf{a}|}{\sin A} = \dfrac{|\mathbf{b}|}{\sin B} = \dfrac{|\mathbf{c}|}{\sin C}.$

WORKED EXAMPLE 53

The position vectors of the points A, B, C are respectively

$\overrightarrow{OA} = \mathbf{a} = -2\mathbf{i} - 3\mathbf{j} + \mathbf{k}$

$\overrightarrow{OB} = \mathbf{b} = 5\mathbf{i} + 2\mathbf{j} - 2\mathbf{k}$

$\overrightarrow{OC} = \mathbf{c} = 3\mathbf{i} - \mathbf{j} - 2\mathbf{k}.$

Find (a) $\overrightarrow{AB} \times \overrightarrow{AC}$ (b) $\overrightarrow{AB} \cdot \overrightarrow{AC}$ (c) $\sin A$ (d) the area of $\triangle ABC$, (e) $\cos A$, (f) the unit vector perpendicular to both $\overrightarrow{AB}$ and $\overrightarrow{AC}$.

Solution 53

(a) $\overrightarrow{AB} = \overrightarrow{OB} - \overrightarrow{OA}$

$= (5\mathbf{i} + 2\mathbf{j} - 2\mathbf{k}) - (-2\mathbf{i} - 3\mathbf{j} + \mathbf{k})$

$= 7\mathbf{i} + 5\mathbf{j} - 3\mathbf{k}$

$\overrightarrow{AC} = \overrightarrow{OC} - \overrightarrow{OA}$

$= (3\mathbf{i} - \mathbf{j} - 2\mathbf{k}) - (-2\mathbf{i} - 3\mathbf{j} + \mathbf{k})$

$= 5\mathbf{i} + 2\mathbf{j} - 3\mathbf{k}$

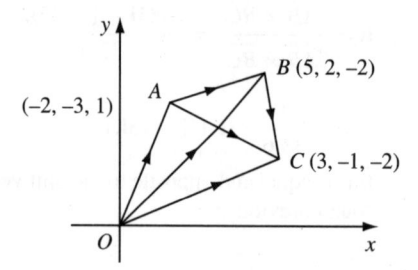

Fig. 8-I/65

$\overrightarrow{AB} \times \overrightarrow{AC} = \begin{vmatrix} \mathbf{i} & \mathbf{j} & \mathbf{k} \\ 7 & 5 & -3 \\ 5 & 2 & -3 \end{vmatrix}$

$= \mathbf{i}(-15 + 6) - \mathbf{j}(-21 + 15) + \mathbf{k}(14 - 25)$

$= -9\mathbf{i} + 6\mathbf{j} - 11\mathbf{k}$

(b) $\overrightarrow{AB} \cdot \overrightarrow{AC} = (7\mathbf{i} + 5\mathbf{j} - 3\mathbf{k}) \cdot (5\mathbf{i} + 2\mathbf{j} - 3\mathbf{k})$

$= 35 + 10 + 9 = 54.$

(c) $\mathbf{a} \times \mathbf{b} = (|\mathbf{a}||\mathbf{b}|\sin\theta)\hat{\mathbf{n}}$

$|\mathbf{a} \times \mathbf{b}| = |\mathbf{a}||\mathbf{b}|\sin\theta|\hat{\mathbf{n}}|$

$|\overrightarrow{AB} \times \overrightarrow{AC}| = |7\mathbf{i}+5\mathbf{j}-3\mathbf{k}||5\mathbf{i}+2\mathbf{j}-3\mathbf{k}|\sin A\,(1)$

$|-9\mathbf{i}+6\mathbf{j}-11\mathbf{k}|$
$= \sqrt{7^2+5^2+3^2}\sqrt{5^2+2^2+3^2}\sin A$

$\sqrt{9^2+6^2+11^2} = \sqrt{83}\sqrt{38}\sin A$
$= \sqrt{238} = 15.43$

$\sin A = \dfrac{15.43}{9.11 \times 6.16} = 0.275$

(d) Area of the triangle $ABC = \dfrac{1}{2}\left|\left(\overrightarrow{AB} \times \overrightarrow{AC}\right)\right|$
$= \dfrac{1}{2}15.43 = 7.72$ square units.

(e) $\overrightarrow{AB} \cdot \overrightarrow{AC} = |\overrightarrow{AB}||\overrightarrow{AC}|\cos A$

$54 = \sqrt{83}\sqrt{38}\cos A$

$\cos A = \dfrac{54}{56} = 0.962$

(f) $\hat{\mathbf{n}} = \dfrac{-9\mathbf{i}+6\mathbf{j}-11\mathbf{k}}{15.43}$

$= -\dfrac{9}{15.43}\mathbf{i} + \dfrac{6}{15.43}\mathbf{j} - \dfrac{11}{15.43}\mathbf{k}$

$\hat{\mathbf{n}} = -0.583\mathbf{i} + 0.389\mathbf{j} - 0.713\mathbf{k}.$

Volume of a Tetrahedron

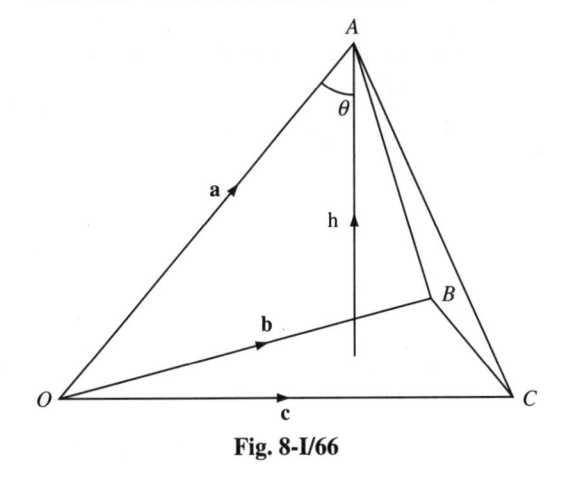

Fig. 8-I/66

Referring to a fixed point O, the position vectors of the point A, B and C are $\mathbf{a}$, $\mathbf{b}$ and $\mathbf{c}$ respectively.

Volume of a tetrahedron
$= \dfrac{1}{3}(\text{area of base}) \times \text{height}$
$= \dfrac{1}{3}\left(\dfrac{1}{2}\text{ area of }\triangle OBC\right) \times \text{height}$
$= \dfrac{1}{6}|\mathbf{b} \times \mathbf{c}||\mathbf{a}|\cos\theta$
$= \dfrac{1}{6}|(\mathbf{b} \times \mathbf{c}) \cdot \mathbf{a}|$

WORKED EXAMPLE 54

The position vectors of the points A, B and C are $\mathbf{a} = -2\mathbf{i}+3\mathbf{j}-5\mathbf{k}$, $\mathbf{b} = \mathbf{i}+4\mathbf{j}+7\mathbf{k}$ and $\mathbf{c} = 3\mathbf{i}-2\mathbf{j}+2\mathbf{k}$. Determine the volume of the tetrahedron with the base $\triangle OBC$ and vertex A.

Solution 54

$V = \dfrac{1}{6}|(\mathbf{b} \times \mathbf{c}) \cdot \mathbf{a}|$

$\mathbf{b} \times \mathbf{c} = \begin{vmatrix} \mathbf{i} & \mathbf{j} & \mathbf{k} \\ 1 & 4 & 7 \\ 3 & -2 & 2 \end{vmatrix}$

$= \mathbf{i}(8+14) - \mathbf{j}(2-21) + \mathbf{k}(-2-12)$
$= 22\mathbf{i} + 19\mathbf{j} - 14\mathbf{k}$

$V = \dfrac{1}{6}|(22\mathbf{i}+19\mathbf{j}-14\mathbf{k}) \cdot (-2\mathbf{i}+3\mathbf{j}-5\mathbf{k})|$

$V = \dfrac{1}{6}|-44+57+70|$

$V = \dfrac{1}{6}(83) = 13.8$ cubic units.

Volume of a Parallelepiped

$\cos\theta = \dfrac{h}{\mathbf{a}}$

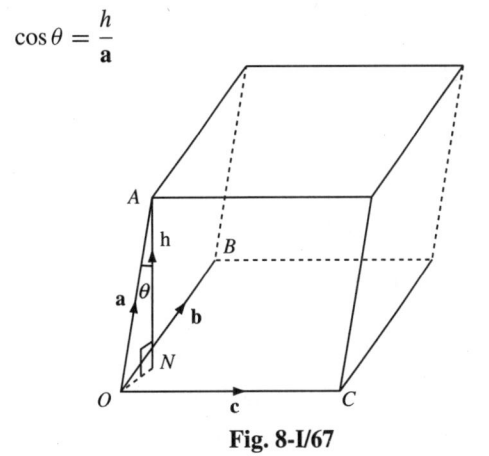

Fig. 8-I/67

The position vectors of the points A, B and C with reference to a fixed point O, the origin, are $\mathbf{a}$, $\mathbf{b}$ and $\mathbf{c}$ respectively.

The volume of a parallelepiped
$$= \text{area of base} \times \text{height}$$
$$= |\mathbf{b} \times \mathbf{c}|h = |\mathbf{b} \times \mathbf{c}||\mathbf{a}|\cos\theta$$

where θ is the angle between $\mathbf{a}$ and h

$$\boxed{V = |(\mathbf{b} \times \mathbf{c}) \cdot \mathbf{a}|}$$

WORKED EXAMPLE 55

Find the volume of a parallelepiped where O is the origin and A, B and C are the points $(-3, 2, 4)$, $(2, -3, -1)$ and $(3, -1, 2)$ respectively.

Solution 55

$\mathbf{a} = -3\mathbf{i} + 2\mathbf{j} + 4\mathbf{k}$,
$\mathbf{b} = 2\mathbf{i} - 3\mathbf{j} - \mathbf{k}$,
$\mathbf{c} = 3\mathbf{i} - \mathbf{j} + 2\mathbf{k}$ $V = |(\mathbf{b} \times \mathbf{c}) \cdot \mathbf{a}|$

$$\mathbf{b} \times \mathbf{c} = \begin{vmatrix} \mathbf{i} & \mathbf{j} & \mathbf{k} \\ 2 & -3 & -1 \\ 3 & -1 & 2 \end{vmatrix}$$

$$= \mathbf{i}(-6 - 1) - \mathbf{j}(4 + 3) + \mathbf{k}(-2 + 9)$$
$$= -7\mathbf{i} - 7\mathbf{j} + 7\mathbf{k}.$$

$\mathbf{b} \times \mathbf{c} \cdot \mathbf{a} = (-7\mathbf{i} - 7\mathbf{j} + 7\mathbf{k}) \cdot (-3\mathbf{i} + 2\mathbf{j} + 4\mathbf{k})$

$$= 21 - 14 + 28 = 35$$

$V = |\mathbf{b} \times \mathbf{c} \cdot \mathbf{a}| = 35$ cubic units.

Alternatively

$V = |(\mathbf{a} \times \mathbf{b}) \cdot \mathbf{c}|$

$$\mathbf{a} \times \mathbf{b} = \begin{vmatrix} \mathbf{i} & \mathbf{j} & \mathbf{k} \\ -3 & 2 & 4 \\ 2 & -3 & -1 \end{vmatrix}$$

$$= \mathbf{i}(-2 + 12) - \mathbf{j}(3 - 8) + \mathbf{k}(9 - 4)$$
$$= 10\mathbf{i} + 5\mathbf{j} + 5\mathbf{k}$$

$\mathbf{a} \times \mathbf{b} \cdot \mathbf{c} = (10\mathbf{i} + 5\mathbf{j} + 5\mathbf{k}) \cdot (3\mathbf{i} - \mathbf{j} + 2\mathbf{k})$

$$= 30 - 5 + 10 = 35 \text{ cubic units.}$$

therefore $\mathbf{b} \times \mathbf{c} \cdot \mathbf{a} = \mathbf{a} \times \mathbf{b} \cdot \mathbf{c}$.

Triple Scalar Product

$\mathbf{a} \times \mathbf{b} \cdot \mathbf{c}$ is called a triple scalar product.

The volume of a parallelepiped $= \mathbf{a} \times \mathbf{b} \cdot \mathbf{c}$ or $= \mathbf{b} \times \mathbf{c} \cdot \mathbf{a}$
therefore $\mathbf{a} \times \mathbf{b} \cdot \mathbf{c} = \mathbf{b} \times \mathbf{c} \cdot \mathbf{a}$.

Note: Do the cross product first and then dot product.
$\mathbf{b} \times \mathbf{c} \cdot \mathbf{a} = \mathbf{a} \cdot \mathbf{b} \times \mathbf{c} = -\mathbf{b} \times \mathbf{a} \cdot \mathbf{c}$.

WORKED EXAMPLE 56

Show that the following pairs of lines are skew:

(i) $\mathbf{r} = \mathbf{i} + \mathbf{k} + \lambda(\mathbf{i} + 3\mathbf{j} + 4\mathbf{k})$
$\mathbf{r} = 2\mathbf{i} + 3\mathbf{j} + \mu(4\mathbf{i} - \mathbf{j} + \mathbf{k})$

(ii) $\mathbf{r} = \mathbf{i} + \mathbf{j} + \lambda(2\mathbf{i} - \mathbf{j} + \mathbf{k})$
$\mathbf{r} = 2\mathbf{i} + \mathbf{j} - \mathbf{k} + \mu(3\mathbf{i} - 5\mathbf{j} + 2\mathbf{k})$

Solution 56

(i) $(a_1 - a_2) \cdot b_1 \times b_2$
$= (\mathbf{i} + \mathbf{k} - 2\mathbf{i} - 3\mathbf{j}) \cdot (\mathbf{i} + 3\mathbf{j} + 4\mathbf{k}) \times (4\mathbf{i} - \mathbf{j} + \mathbf{k})$
$= (-\mathbf{i} - 3\mathbf{j} + \mathbf{k}) \cdot (7\mathbf{i} + 15\mathbf{j} - 13\mathbf{k})$
$= -7 - 45 - 13 = -65$

$(\mathbf{i} + 3\mathbf{j} + 4\mathbf{k}) \times (4\mathbf{i} - \mathbf{j} + \mathbf{k})$

$$= \begin{vmatrix} \mathbf{i} & \mathbf{j} & \mathbf{k} \\ 1 & 3 & 4 \\ 4 & -1 & 1 \end{vmatrix}$$

$$= \mathbf{i}(3 + 4) - \mathbf{j}(1 - 16) + \mathbf{k}(-1 - 12)$$
$$= 7\mathbf{i} + 15\mathbf{j} - 13\mathbf{k}$$

the lines do not intersect since $(a_1 - a_2) \cdot b_1 \times b_2$ is not equal to zero and the lines are not parallel since the direction ratios of the lines are different $1 : 3 : 4$ and $4 : -1 : 1$. Therefore the lines are skew.

(ii) $(a_1 - a_2) \cdot b_1 \times b_2 = (\mathbf{i} + \mathbf{j} - 2\mathbf{i} - \mathbf{j} + \mathbf{k}) \cdot b_1 \times b_2$
$= (-\mathbf{i} + \mathbf{k}) \cdot b_1 \times b_2$
$b_1 \times b_2 = (2\mathbf{i} - \mathbf{j} + \mathbf{k}) \times (3\mathbf{i} - 5\mathbf{j} + 2\mathbf{k})$

$$= \begin{vmatrix} \mathbf{i} & \mathbf{j} & \mathbf{k} \\ 2 & -1 & 1 \\ 3 & -5 & 2 \end{vmatrix}$$

$$= \mathbf{i}(-2 + 5) - \mathbf{j}(4 - 3) + \mathbf{k}(-10 + 3)$$
$$= 3\mathbf{i} - \mathbf{j} - 7\mathbf{k}$$

$(a_1 - a_2) \cdot b_1 \times b_2 = (-\mathbf{i} + \mathbf{k}) \cdot (3\mathbf{i} - \mathbf{j} - 7\mathbf{k})$
$$= -3 - 7 = -10$$

the lines do not intersect.

$x = 1 + 2\lambda \quad y = 1 - \lambda$

$\dfrac{x-1}{2} = \dfrac{y-1}{1} = \dfrac{z}{1} = \lambda$

$z = \lambda \quad x = 2 + 3\mu$

$y = 1 - 5\mu \quad \dfrac{x-2}{3} = \dfrac{y-1}{-5} = \dfrac{z+1}{2} = \mu$

$z = -1 + 2\mu$

$2 : -1 : 1, 3 : -5 : 2$ the direction ratios are not the same, therefore the lines are not parallel. The lines are skew.

Consider the example.

Worked Example 57

Find the perpendicular distance of a point $C(1, 2, 3)$ from the line with a vector equation $\mathbf{r} = (2\mathbf{i} + 3\mathbf{j} + 4\mathbf{k}) + t(-3\mathbf{i} + 4\mathbf{j} - \mathbf{k})$.

Solution 57

$d = \dfrac{|\mathbf{b} \times (\mathbf{c} - \mathbf{a})|}{|\mathbf{b}|}$

$\mathbf{b} = -3\mathbf{i} + 4\mathbf{j} - \mathbf{k}$ the direction vector

$\mathbf{a} = 2\mathbf{i} + 3\mathbf{j} + 4\mathbf{k}$ the position vector

$\mathbf{c} = \mathbf{i} + 2\mathbf{j} + 3\mathbf{k}$ the position vector of the point C

$|\mathbf{b}| = \sqrt{3^2 + 4^2 + 1^2} = \sqrt{26}$

$\mathbf{c} - \mathbf{a} = \mathbf{i} + 2\mathbf{j} + 3\mathbf{k} - 2\mathbf{i} - 3\mathbf{j} - 4\mathbf{k} = -\mathbf{i} - \mathbf{j} - \mathbf{k}$

$\mathbf{b} \times (\mathbf{c} - \mathbf{a}) = (-3\mathbf{i} + 4\mathbf{j} - \mathbf{k}) \times (-\mathbf{i} - \mathbf{j} - \mathbf{k})$

$= \begin{vmatrix} \mathbf{i} & \mathbf{j} & \mathbf{k} \\ -3 & 4 & -1 \\ -1 & -1 & -1 \end{vmatrix}$

$= \mathbf{i}(-4 - 1) - \mathbf{j}(3 - 1) + \mathbf{k}(3 + 4)$

$= -5\mathbf{i} - 2\mathbf{j} + 7\mathbf{k}$

$d = \dfrac{|-5\mathbf{i} - 2\mathbf{j} + 7\mathbf{k}|}{\sqrt{26}}$

$= \dfrac{\sqrt{25 + 4 + 49}}{\sqrt{26}} = \sqrt{3}$

The Shortest Distance between Two Skew Lines

Let the vector equations of two skew lines be

$l_1 : \mathbf{r}_1 = \mathbf{a}_1 + \lambda \mathbf{b}_1$

$l_2 : \mathbf{r}_2 = \mathbf{a}_2 + \mu \mathbf{b}_2$

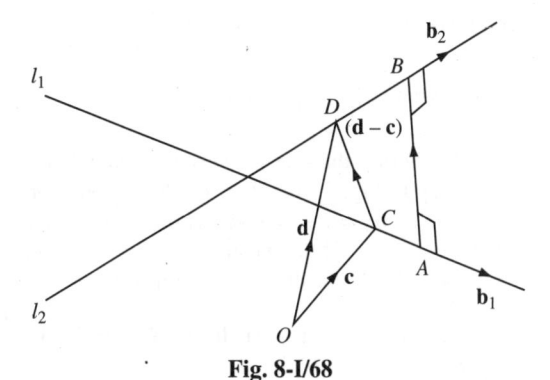

Fig. 8-I/68

The shortest distance between the two lines must be perpendicular to the lines, let this be $\overrightarrow{AB}$ as shown, the cross product of the positive direction vectors

$\mathbf{b}_1 \times \mathbf{b}_2$ is parallel to $\overrightarrow{AB}$.

The unit vector, $\hat{\mathbf{n}}$, in the direction of $\overrightarrow{AB}$ is $\dfrac{\mathbf{b}_1 \times \mathbf{b}_2}{|\mathbf{b}_1 \times \mathbf{b}_2|}$.

If the angle between $\overrightarrow{CD}$ and $\overrightarrow{AB}$ is θ

$\left|\overrightarrow{AB}\right| = \left|\overrightarrow{CD}\right| \cos\theta = \left|\overrightarrow{CD} \cdot \hat{\mathbf{n}}\right|$

since $\left|\overrightarrow{CD}\right| \cdot \hat{\mathbf{n}} = \left|\overrightarrow{CD}\right| |\hat{\mathbf{n}}| \cos\theta$ where $|\hat{\mathbf{n}}| = 1$

Therefore, $\left|\overrightarrow{AB}\right| = \left|\overrightarrow{CD} \cdot \hat{\mathbf{n}}\right|$ and $\overrightarrow{CD} = \mathbf{d} - \mathbf{c}$.

$$\boxed{d = \left|(\mathbf{d} - \mathbf{c}) \cdot \dfrac{\mathbf{b}_1 \times \mathbf{b}_2}{|\mathbf{b}_1 \times \mathbf{b}_2|}\right|}$$

the shortest distance between two skew lines. If the lines intersect then $d = 0$

$$\boxed{(\mathbf{d} - \mathbf{c}) \cdot \mathbf{b}_1 \times \mathbf{b}_2 = 0}$$

Worked Example 58

Find the shortest distance between the two skew lines.

(i) $l_1 : \mathbf{r} = \mathbf{i} + \mathbf{k} + \lambda(\mathbf{i} + 3\mathbf{j} + 4\mathbf{k})$
    $l_2 : \mathbf{r} = 2\mathbf{i} + 3\mathbf{j} + \mu(4\mathbf{i} - \mathbf{j} + \mathbf{k})$

(ii) $l_1 : \mathbf{r} = \mathbf{i} + \mathbf{j} + \lambda(2\mathbf{i} - \mathbf{j} + \mathbf{k})$
    $l_2 : \mathbf{r} = 2\mathbf{i} + \mathbf{j} - \mathbf{k} + \mu(3\mathbf{i} - 5\mathbf{j} + 2\mathbf{k})$

(iii) $l_1 : \mathbf{r} = \mathbf{i} - 2\mathbf{j} + 3\mathbf{k} + \lambda(-\mathbf{i} + \mathbf{j} - 2\mathbf{k})$
$l_2 : \mathbf{r} = \mathbf{i} - \mathbf{j} - \mathbf{k} + \mu(\mathbf{i} + 2\mathbf{j} - 2\mathbf{k})$

(iv) $l_1 : \mathbf{r} = -3\mathbf{i} + \mathbf{j} + \lambda(2\mathbf{i} + \mathbf{j} + 2\mathbf{k})$
$l_2 : \mathbf{r} = 8\mathbf{i} + 3\mathbf{j} + 15\mathbf{k} + \mu(3\mathbf{i} + 2\mathbf{j} + 5\mathbf{k})$

Solution 58

(i) $d = \left| (\mathbf{d} - \mathbf{c}) \cdot \dfrac{\mathbf{b}_1 \times \mathbf{b}_2}{|\mathbf{b}_1 \times \mathbf{b}_2|} \right|$ where

d is the shortest distance between two skew lines. Referring to Fig. 8-I/68, $\mathbf{d} = \mathbf{i} + \mathbf{k}$, the position vector of l_1, $\mathbf{b}_2 = \mathbf{i} + 3\mathbf{j} + 4\mathbf{k}$ the direction vector of l_1; $\mathbf{c} = 2\mathbf{i} + 3\mathbf{j}$, the position vector of l_2 and $\mathbf{b}_1 = 4\mathbf{i} - \mathbf{j} + \mathbf{k}$, the direction vector of l_2.

$\mathbf{d} = \mathbf{i} + \mathbf{k}, \mathbf{c} = 2\mathbf{i} + 3\mathbf{j}, \mathbf{b}_2 = \mathbf{i} + 3\mathbf{j} + 4\mathbf{k}.$
$\mathbf{b}_1 = 4\mathbf{i} - \mathbf{j} + \mathbf{k}$

$d = \left| (\mathbf{i} + \mathbf{k} - 2\mathbf{i} - 3\mathbf{j}) \cdot \dfrac{(4\mathbf{i} - \mathbf{j} + \mathbf{k}) \times (\mathbf{i} + 3\mathbf{j} + 4\mathbf{k})}{|(4\mathbf{i} - \mathbf{j} + \mathbf{k}) \times (\mathbf{i} + 3\mathbf{j} + 4\mathbf{k})|} \right|$

$(4\mathbf{i} - \mathbf{j} + \mathbf{k}) \times (\mathbf{i} + 3\mathbf{j} + 4\mathbf{k})$

$= \begin{vmatrix} \mathbf{i} & \mathbf{j} & \mathbf{k} \\ 4 & -1 & 1 \\ 1 & 3 & 4 \end{vmatrix}$

$= \mathbf{i}(-4 - 3) - \mathbf{j}(16 - 1) + \mathbf{k}(12 + 1)$

$= -7\mathbf{i} - 15\mathbf{j} + 13\mathbf{k}.$

$|(4\mathbf{i} - \mathbf{j} + \mathbf{k}) \times (\mathbf{i} + 3\mathbf{j} + 4\mathbf{k})|$

$= |-7\mathbf{i} - 15\mathbf{j} + 13\mathbf{k}|$

$= \sqrt{49 + 225 + 169}$

$= 21.1$

$d = \left| \dfrac{(-\mathbf{i} - 3\mathbf{j} + \mathbf{k}) \cdot (-7\mathbf{i} - 15\mathbf{j} + 13\mathbf{k})}{21.1} \right|$

$= \dfrac{7 + 45 + 13}{21.1}$

$= 3.08 \text{ units}$

(ii) $\mathbf{d} = \mathbf{i} + \mathbf{j}, \mathbf{c} = 2\mathbf{i} + \mathbf{j} - \mathbf{k}, \mathbf{b}_2 = 2\mathbf{i} - \mathbf{j} + \mathbf{k},$
$\mathbf{b}_1 = (3\mathbf{i} - 5\mathbf{j} + 2\mathbf{k})$

$d = \left| (\mathbf{d} - \mathbf{c}) \cdot \dfrac{\mathbf{b}_1 \times \mathbf{b}_2}{|\mathbf{b}_1 \times \mathbf{b}_2|} \right|$

$\mathbf{b}_1 \times \mathbf{b}_2 = \begin{vmatrix} \mathbf{i} & \mathbf{j} & \mathbf{k} \\ 3 & -5 & 2 \\ 2 & -1 & 1 \end{vmatrix}$

$= \mathbf{i}(-5 + 2) - \mathbf{j}(3 - 4) + \mathbf{k}(-3 + 10)$

$= -3\mathbf{i} + \mathbf{j} + 7\mathbf{k}$

$|\mathbf{b}_1 \times \mathbf{b}_2| = |-3\mathbf{i} + \mathbf{j} + 7\mathbf{k}|$

$= \sqrt{9 + 1 + 49}$

$= \sqrt{59}$

$= 7.68.$

$d = \left| \dfrac{(-\mathbf{i} + \mathbf{k}) \cdot (-3\mathbf{i} + \mathbf{j} + 7\mathbf{k})}{7.68} \right|$

$= \dfrac{3 + 7}{7.58} = \dfrac{10}{7.58}$

$= 1.30$

(iii) $\mathbf{d} = \mathbf{i} - 2\mathbf{j} + 3\mathbf{k}, \mathbf{c} = \mathbf{i} - \mathbf{j} - \mathbf{k}, \mathbf{b}_2 = -\mathbf{i} + \mathbf{j} - 2\mathbf{k},$
$\mathbf{b}_1 = \mathbf{i} + 2\mathbf{j} - 2\mathbf{k}$

$\mathbf{d} - \mathbf{c} = \mathbf{i} - 2\mathbf{j} + 3\mathbf{k} - \mathbf{i} + \mathbf{j} + \mathbf{k} = -\mathbf{j} + 4\mathbf{k}$

$\mathbf{b}_1 \times \mathbf{b}_2 = \begin{vmatrix} \mathbf{i} & \mathbf{j} & \mathbf{k} \\ 1 & 2 & -2 \\ -1 & 1 & -2 \end{vmatrix}$

$= \mathbf{i}(-4 + 2) - \mathbf{j}(-2 - 2) + \mathbf{k}(1 + 2)$

$= -2\mathbf{i} + 4\mathbf{j} + 3\mathbf{k}$

$|\mathbf{b}_1 \times \mathbf{b}_2| = |-2\mathbf{i} + 4\mathbf{j} + 3\mathbf{k}|$

$= \sqrt{4 + 16 + 9}$

$= \sqrt{29}$

$d = \left| \dfrac{(-\mathbf{j} + 4\mathbf{k}) \cdot (-2\mathbf{i} + 4\mathbf{j} + 3\mathbf{k})}{\sqrt{29}} \right|$

$= \dfrac{-4 + 12}{\sqrt{29}}$

$= \dfrac{8}{\sqrt{29}}$

$= 1.49$

(iv) $\mathbf{d} = -3\mathbf{i} + \mathbf{j}, \mathbf{c} = 8\mathbf{i} + 3\mathbf{j} + 15\mathbf{k}, \mathbf{b}_2 = 2\mathbf{i} + \mathbf{j} + 2\mathbf{k},$
$\mathbf{b}_1 = 3\mathbf{i} + 2\mathbf{j} + 5\mathbf{k}$

$\mathbf{d} - \mathbf{c} = -3\mathbf{i} + \mathbf{j} - 8\mathbf{i} - 3\mathbf{j} - 15\mathbf{k} = -11\mathbf{i} - 2\mathbf{j} - 15\mathbf{k}$

$$\mathbf{b}_1 \times \mathbf{b}_2 = \begin{vmatrix} \mathbf{i} & \mathbf{j} & \mathbf{k} \\ 3 & 2 & 5 \\ 2 & 1 & 2 \end{vmatrix}$$

$$= \mathbf{i}(4-5) - \mathbf{j}(6-10) + \mathbf{k}(3-4)$$

$$= -\mathbf{i} + 4\mathbf{j} - \mathbf{k}$$

$$|\mathbf{b}_1 \times \mathbf{b}_2| = |-\mathbf{i} + 4\mathbf{j} - \mathbf{k}|$$

$$= \sqrt{1 + 16 + 1} = \sqrt{18} = 4.24$$

$$\mathbf{d} = \left| \frac{(-11\mathbf{i} - 2\mathbf{j} - 15\mathbf{k}) \cdot (-\mathbf{i} + 4\mathbf{j} - \mathbf{k})}{4.24} \right|$$

$$= \frac{11 - 8 + 15}{4.24}$$

$$= \frac{18}{4.24} = 4.25.$$

Formulae (Summary) Planes

$$\boxed{ax + by + cz = d}$$

the cartesian equation of a plane where $a : b : c$ are the direction ratios of a normal to the plane

$$\boxed{\frac{ax_1 + by_1 + cz_1 - d}{\sqrt{a^2 + b^2 + c^2}}}$$

the perpendicular distance of the point $A(x_1, y_1, z_1)$ from the plane $ax + by + cz = d$, where the direction cosines are $l : m : n$, or $\frac{a}{\sqrt{a^2 + b^2 + c^2}} : \frac{b}{\sqrt{a^2 + b^2 + c^2}} : \frac{c}{\sqrt{a^2 + b^2 + c^2}}$, or $\cos \alpha : \cos \beta : \cos \gamma$.

$$\boxed{\mathbf{a} \times \mathbf{b} = (|\mathbf{a}||\mathbf{b}| \sin \theta)\hat{\mathbf{n}}}$$

the vector product of $\mathbf{a}$ and $\mathbf{b}$ in a direction perpendicular to the plane containing $\mathbf{a}$ and $\mathbf{b}$ in the sense a right-handed screw turned from $\mathbf{a}$ to $\mathbf{b}$.

$$\boxed{\mathbf{a} \times \mathbf{b} = -\mathbf{b} \times \mathbf{a}}$$

the vector product is <u>not</u> commutative.

$$\mathbf{a} \times \mathbf{b} = \begin{vmatrix} \mathbf{i} & \mathbf{j} & \mathbf{k} \\ x_1 & y_1 & z_1 \\ x_2 & y_2 & z_2 \end{vmatrix}$$

$$\mathbf{a} = x_1 \mathbf{i} + y_1 \mathbf{j} + z_1 \mathbf{k},$$
$$\mathbf{b} = x_2 \mathbf{i} + y_2 \mathbf{j} + z_2 \mathbf{k}.$$

$$\boxed{|\mathbf{a} \times \mathbf{b}| = |\mathbf{a}||\mathbf{b}| \sin \theta}$$

$$\boxed{\text{Area of a Triangle } ABC = \frac{1}{2}|\mathbf{a} \times \mathbf{b}|} \text{ where } \mathbf{a}$$
and $\mathbf{b}$ are two sides.

$$\boxed{\mathbf{r} \cdot \mathbf{n} = d}$$ the standard form of the vector equation of a plane.

$$\boxed{p = \mathbf{a} \cdot \hat{\mathbf{n}} - D}$$

perpendicular distance of a point from a plane, the point P and the origin O are on opposite side of the plane Π.

$$\boxed{\cos \theta = \hat{\mathbf{n}}_1 \cdot \hat{\mathbf{n}}_2}$$

the angle between two planes Π_1 and Π_2.

$$\boxed{\hat{\mathbf{n}}_1 = \hat{\mathbf{n}}_2}$$

the planes Π_1 and Π_2 are parallel.

$$\boxed{\hat{\mathbf{n}}_1 \cdot \hat{\mathbf{n}}_2 = 0}$$

the planes Π_1 and Π_2 are perpendicular.

$$\boxed{\sin \theta = \frac{\mathbf{b} \cdot \hat{\mathbf{n}}}{|\mathbf{b}|}}$$

the angle between a line and a plane Π, where $\mathbf{b}$ is the direction vector of a line.

$$\boxed{\mathbf{r} \cdot (\hat{\mathbf{n}}_1 - k\hat{\mathbf{n}}_2) = D_1 - kD_2}$$

the equation of the plane passing through the intersections of two planes Π_1 and Π_2 with vector equations $\mathbf{r} \cdot \hat{\mathbf{n}}_1 = D_1$ and $\mathbf{r} \cdot \hat{\mathbf{n}}_2 = D_2$.

$$\boxed{\mathbf{r} = \mathbf{a} + s\mathbf{b} + t\mathbf{c}}$$

the vector equation of a plane through the point with position vector $\mathbf{a}$ and parallel to $\mathbf{b}$ and $\mathbf{c}$.

Volume of a tetrahedron $\boxed{V = \frac{1}{6}|(\mathbf{b} \times \mathbf{c}) \cdot \mathbf{a}|}$

where $\mathbf{b}$ and $\mathbf{c}$ are any two sides of the base and $\mathbf{a}$ is the position vector of the vertex. Alternatively

$$\boxed{V = \frac{1}{6}|\mathbf{a} \times \mathbf{b} \cdot \mathbf{c}|}$$

Volume of a parallelepiped $\boxed{V = |\mathbf{b} \times \mathbf{c} \cdot \mathbf{a}| = |\mathbf{a} \times \mathbf{b} \cdot \mathbf{c}|}$

Condition that two lines intersect with vector equations $\mathbf{r}_1 = \mathbf{a}_1 + \lambda \mathbf{b}_1$, $\mathbf{r}_2 = \mathbf{a}_2 + \mu \mathbf{b}_2$

$$\boxed{(\mathbf{a}_1 - \mathbf{a}_2) \cdot \mathbf{b}_1 \times \mathbf{b}_2 = 0}$$

Exercises 6

1. Verify the anti commutative property for the vectors

   $u = 3i + 3j + 5k$
   $v = -2i + 4j + k$

   (i) $u \times v$ and (ii) $v \times u$.

2. Verify the anti commutative property for the following pairs of vectors:

   (a) $u = i + j + k \quad v = -2i - j + k$

   $u \times v$ and $v \times u$

   (b) $u = \begin{pmatrix} 3 \\ -5 \\ 3 \end{pmatrix}, \quad v = \begin{pmatrix} -5 \\ 3 \\ 1 \end{pmatrix}$

   $u \times v$ and $v \times u$.

   (c) $v = \begin{pmatrix} -3 \\ -4 \\ +5 \end{pmatrix}, \quad w = \begin{pmatrix} 1 \\ 1 \\ 1 \end{pmatrix}$

   $v \times w$ and $w \times v$.

3. Show that

   (i) $(a \times b) \cdot a$
   (ii) $(a \times b) \cdot b$

   are orthogonal.

4. Show that a is parallel to b if and only if $a \times b = 0$.

5. Find $u \times v$.

   (i) $u = i + 2j + 3k, \quad v = -i + 2j + k$
   (ii) $u = 4i + 3k, \quad v = 3j - k$
   (iii) $u = 2i + j + 2k, \quad v = i - j + 2k$.

6. Show that

   (i) $i \times j = k$
   (ii) $i \times k = -j$
   (iii) $j \times k = i$
   (iv) $k \times i = j$
   (v) $i \times j \times k = j \times k \times i = k \times i \times j$.

7. Find an equation of the plane through the three points P, Q and R given by the position vectors $p = i + 2j - 5k, q = -3i + 4j - k, r = 3i + 5k$.

8. Find an equation of the plane through the three points P, Q and R given by the position vectors $p = i + 2j - 5k, q = -3i + 4j - k, r = 3i + 5k$.

9. Find the perpendicular distance of a point $A(4, 5, 6)$ from the line with a vector equation $r = i + j + k + \lambda(-2i + 3j - k)$.

10. Find the perpendicular distance of a point $A(2, -1, 3)$ from the line with a vector equation $r = -2i + j - k + \lambda(5i - j - k)$.

11. Prove that the perpendicular distance of a point $C(x_1, y_1, z_1)$ from the line with a vector equation $r = a + \lambda b$ is given by $d = \dfrac{|b \times (c - a)|}{|b|}$.

12. Show that the area of $\triangle ABC$ is given $\dfrac{1}{2}|a \times b|$.

13. Show that the volume of a tetrahedron is given by $V = \dfrac{1}{6}|(b \times c) \cdot a|$ where b and c are any two sides of the base and a is the position vector or the vertex.

14. Show that the volume of a parallelipid is given by $V = |(b \times c) \cdot a|$ where a, b and c are the position vectors.

15. The position vectors of the vertices of a tetrahedron $ABCD$ given as follows:

    $a = i - 2j + 6k$
    $b = 2i + 3j - k$
    $c = 5i - 4j - 3k$
    $d = -3i + 2j + 2k$

    (i) Determine the angle between the faces ABC and BCD.

    (ii) Determine the angle between the faces ACD and ABD.

16. The position vector of the points P, Q and R are $2i - 3j + 2k, -3i + j + 4k$, and $3i + 2j - 3k$ respectively.

    (a) Find $\overrightarrow{PQ} \times \overrightarrow{PR}$.

    (b) Hence calculate the area of $\triangle PQR$.

    (c) Determine the equation of the plane PQR in the form $r \cdot n = d$.

17. The position vectors of the points A, B and C are $4j - 5k, 2i - j + k$, and $-3i + j + 7k$ respectively

(a) Find $\overrightarrow{AB} \times \overrightarrow{AC}$.

(b) Hence calculate the area of $\triangle ABC$.

(c) Determine the equation of the plane ABC in the form $\mathbf{r} \cdot \mathbf{n} = d$.

18. Show from first principles that the vector product $\mathbf{a} \times \mathbf{b}$ is given by the

$$\text{determinant} \begin{vmatrix} \mathbf{i} & \mathbf{j} & \mathbf{k} \\ a_1 & a_2 & a_3 \\ b_1 & b_2 & b_3 \end{vmatrix}.$$

19. Determine the perpendicular distances from the point $P(1, 1, 1)$ to the following lines:

(i) $\mathbf{r} = (2 + 3\lambda)\mathbf{i} + (1 - 2\lambda)\mathbf{j} + (3 + \lambda)\mathbf{k}$

(ii) $\mathbf{r} = (1 - t)\mathbf{i} + (1 + t)\mathbf{j} + (1 - 2t)\mathbf{k}$

(iii) $\mathbf{i} + \mathbf{j} - 3\mathbf{k} + t(-2\mathbf{i} + 3\mathbf{j} + 4\mathbf{k})$.

20. Find the distance of the point $P(-1, -2, -4)$ from the line $\mathbf{r} = -\mathbf{i} + 7\mathbf{k} + \mu(\mathbf{i} - 2\mathbf{j} - 3\mathbf{k})$.

21. Find the distance of the point $P(5, 8, 9)$ from the line $\mathbf{r} = 2\mathbf{i} + 5\mathbf{k} + V(3\mathbf{i} + 4\mathbf{j} + 8\mathbf{k})$.

22. Find the vector equation of the line which passes through the point $P(1, 2, -1)$ and which is perpendicular to the plane containing the vectors $\mathbf{v} = (1, -1, 1)$, $\mathbf{w}(2, 1, 3)$.

8. VECTORS IN TWO AND THREE DIMENSIONS

Index

A
acceleration centripetal   4–5
addition of algebra   2
algebra addition   2
      subtraction   2
angle of triangle   46
angle between a pair of line   22
angle between two planes   34
angle between a line and a plane   35
anti-commutative property of vectors   40

B
bearings   4

C
cardinals   4
cartesian equation of a line   17
centripetal acceleration   5
collinear points   19
column matrix   7
coordinate geometry of
three dimensions   12, 27

D
direction ratios of a vector   9
direction cosines of a vector   9
distance between two parallel planes   31
distance of a plane from the origin   32
distance of a point from a line   50

E
equation of plane   27–32

F
forces   2
formulae (summary planes)   52

I
intersection of a pair of lines   21
intersection of two planes   35

L
line passing through two fixed two points   18

M
magnitude of a vector or modulus of a vector   1

N
negative vector   2

O
orthogonal vector   6–7

P
pairs of lines   21
parallel lines   21
parallel planes   33–4
parametric form of a vector
      equation of a plane   36–7
perpendicular distance
      of a point from a plane   30, 32–3
plane passing through a given point
      and perpendicular to a given direction   32
position vector   2

S
scalar multiple or submultiple of a vector   3
scalar quantity   1
scalar product   23
sine rule   4
shortest distance between two skewed lines   50

T
tetrahedron   48
triangle of forces   2
three collinear points   19
triple scalar   49

U
unit vector   6

V
Vector
      components   7–8
      direction ratios   9
      direction cosines   9–11
      equation of a straight line   15–16

equation or plane   31
free   2
geometrical representation
    magnitude   10–11
negative   2
parallelogram   3
polygon   3

position   2
product   40
scalar multiple or submultiple   3

Velocity   3–4
Volume of a parallel piped   48
Volume of a tetrahedron   53